Plant Layout and Flow Improvement

Jay Cedarleaf

McGraw-Hill, Inc.
New York San Francisco Washington, D.C. Auckland Bogotá
Caracas Lisbon London Madrid Mexico City Milan
Montreal New Delhi San Juan Singapore
Sydney Tokyo Toronto

Library of Congress Cataloging-in-Publication Data

Cedarleaf, Jay.
    Plant layout and flow improvement / Jay Cedarleaf.
        p.   cm.
    Includes bibliographical references and index.
    ISBN 0-07-011043-3
    1. Plant layout. 2. Production planning. I. Title.
    TS178.C33   1994
    658.2'3—dc20   94-4207
                                                                        CIP

    2 3 4 5 6 7 8 9 0   KPT/KPT   9 0 9 8 7 6 5

ISBN 0-07-011043-3

The sponsoring editor for this book was Robert W. Hauserman, the editing supervisor was David E. Fogarty, and the production supervisor was Suzanne W. Babeuf. This book was set in Garamond by McGraw-Hill's Professional Book Group composition unit.

Printed and bound by Arcata Graphics/Kingsport.

The registered trademarks of computer programs, operating systems, and companies that are referenced in this book are listed here. Owners of the trademarks are given in parentheses. AutoCAD (Autodesk, Inc.); Generic CADD (Autodesk, Inc.); Lotus and 1-2-3 (Lotus Development Corp.); WordPerfect (WordPerfect Corp.); CADKEY (Cadkey, Inc.); MS-DOS and Windows (Microsoft Corp.); Quick Schedule Plus (Power-Up! Software Corp.).

Contents

Acknowledgments

To my parents, Margaret and Cliff: Thank you both for giving me free rein, as a boy, to pursue whatever endeavor took my fancy. I cannot remember a single instance when you discouraged me from pursuing my current interest—from laying out a tree house to manufacturing a go-cart.

To my wife, Tricia: Thank you for checking everything in this book and for being my devoted editor. Whenever a statement seemed ambiguous or confusing, you discovered what I really meant and made sure that I said it.

My gratitude to my fellow industrial engineer, Leslie H. Brighton, for reviewing my book proposal and offering many valuable suggestions. He also advised me about how to design the flow arrows to better illustrate the dynamics of the material flow in my plant layouts.

I also wish to thank the following corporations that I have worked for and consulted with. I was able to develop the techniques presented in this book because they allowed me the freedom to investigate new approaches to their layout and manufacturing problems.

Digital Equipment Corporation

Ford Aerospace Division

Hewlett-Packard Corporation

Litton Data Systems Division

Loral Command & Control Systems

Square D Corporation

Telex Corporation

United Technologies Corporation

Jay Cedarleaf

About the Author

Jay Cedarleaf, a graduate industrial engineer, is an experienced consultant specializing in plant layout and product cost analysis. He has helped such Fortune 500 corporations as Digital Equipment, Ford Aerospace, Hewlett-Packard, Litton Data Systems, Loral, Square D, and United Technologies design their plant layouts, manage the move of entire facilities, and develop product cost reduction programs. Mr. Cedarleaf is a senior member of the Institute of Industrial Engineers.

Explanation

Welcome to the field of plant layout. It is a branch of the broader category of facilities layout, which includes large establishments such as banks, stores, and schools. You are entering the center ring of the manufacturing arena, one of the few realms of engineering that is an art, not a science. The importance of this field is stated in the *Handbook of Industrial Engineering:*[1]*

> It has been estimated that between 20% and 50% of the total operating expenses within manufacturing are attributed to material handling. It is generally agreed that effective facilities layout can reduce these costs by at least 10 to 30%. If effective facilities layout were thus applied, the annual manufacturing productivity in the United States would increase approximately three times more than it has in any year in the last decade [p. 1777].
>
> Facilities layout, then, although becoming more scientific, remains today as an art [p. 1778].

As a plant layout designer, you are a technical artist creating a one-of-a-kind drawing of an industrial system for the efficient manufacture of a product. Your design can influence the very future of your company. If production improvements are incorporated into your layout, the results can be dramatic in both the cost and the quality of the finished product.

Although you can use a drafting table and a calculator with this book to complete a project successfully, a computer will substantially enhance both the speed and the quality of your output. The instructions are, therefore, tailored for the computer user. AutoCAD, Lotus 1-2-3, and WordPerfect programs are used to illustrate many of the subjects covered in this book. However, the instructions are written so that any computer-aided design

---

*Superscript numbers refer to entries in the References section at the end of the book.

(CAD), spreadsheet, or word processing program can be used instead.

The Purpose

If you have been assigned a plant layout project and do not know where to begin, this book has been designed for you. Much of this material has never been published before. Even the experienced designer will benefit from the state-of-the-art techniques described in detail in every chapter. It is a complete step-by-step guide. Few designers, including industrial engineers, have the opportunity to plan and implement a major manufacturing layout more than once in their careers. As a result, the whole process must be learned from scratch without the benefit of direct experience. This book provides that experience.

The principles and instructions given here can also be applied to a warehouse, an order-picking business, or any facility where the flow of material is critical to success. The flow analysis strategy described in this book can even be used to streamline the processing of paperwork in an office. Once you have mastered these basic principles, the applications are endless and can have a profound influence on the future of your company, limited only by your imagination.

This book will guide you through the entire layout project, from streamlining the flow before designing the layout to moving into the new plant. Surprisingly, designing the detailed layout of equipment in the new plant is a small part of the total layout project. The investigation, data gathering, and preparation for the layout are far more time-consuming than the layout itself.

Of course, if you have direct experience with any step, you can review that text briefly and just use the checklists as a guide. Because some readers will not study every section, this book is not divided into numbered steps. Rather, it is structured along a series of logical, sequential subjects. No matter what your degree of experience, this book can broaden your background in designing plant layouts.

Common errors can overwhelm the uninitiated who are not familiar with the disasters that can occur during the turmoil of the layout project execution. It is a frustrating experience to realize afterward that the solutions were obvious, if only the conditions had been anticipated. The checklists in this book will help you foresee these conditions while you can still prevent them.

The purpose of this book is to help you:

1. Provide a system for improving the product cost and quality.

2. Organize the layout project into a structured, efficient plan.

3. Anticipate and solve common plant layout problems.

4. Ensure that no vital steps of the process are skipped.

5. Ensure that everyone is well satisfied with the results.

The first few chapters in this book describe some basic concepts that have been used by the author to improve the efficiency of manufacturing systems in the high-tech industry. Generally, *just-in-time* (JIT) principles will guide the improvement project, but with a different approach than is commonly used in manufacturing.

In this book, JIT will not be applied immediately to the assembly process, as this can result in an immense waste of time and money. Instead, the improvements will be quantified and justified before any concrete steps are taken to change the production process. The improvement plan can then be presented to management for consideration in light of the implementation costs versus the well-defined and quantified benefits. The whole process is much more refined than the blind-faith advocacy of JIT, which was characteristic of the original discipline.

Each chapter, which may include several steps, is a well-defined division of the total project, and is written in sufficient detail so that you will understand exactly how to proceed. A chapter can be read in one sitting and the information used immediately. After absorbing what is to be accomplished, you can start a plan of action that will complete those steps successfully.

Checklists have been provided to ensure that no common items will be overlooked. Bear in mind that not all instructions and checklists necessarily apply to every industry or plant. Also, the instructions are for a move from plant to plant, but can apply to a move within a plant as well.

The drawings and analyses used in this book can be done manually, but are described in terms of the following programs:

1. Drawings of flowcharts, pie charts, relationship charts, plant layouts, and the block library are constructed by using AutoCAD.[3]

2. The labor analysis and the equipment list spreadsheet are constructed by using Lotus 1-2-3.[5]

3. The list of AutoCAD blocks and the moving schedule are constructed by using WordPerfect.[6]

AutoCAD commands are capitalized when they are used with the word *command* (use the INSERT command), when the spelling is unidentifiable (WBLOCK it to the hard disk), or when the usage is uncommon (FILL the walls later). Commands are not capitalized when they have a familiar meaning, such as the commands *insert* and *erase* (after you insert the blocks, erase them).

Many of the illustrations in this book are horizontal. That makes them inconvenient to study since you must turn the book to study the figure and then turn it back again to read the text. It is therefore recommended that you make copies of all illustrations and keep them readily available. In that way, you can read the text and examine the illustrations simultaneously without losing your train of thought or your place in the book. Also, you can easily refer to the illustrations while you are at the computer or out on the plant floor.

Of necessity, the information density of this book is very high. For a project of this nature, time is of the essence, so each sentence is simple and straightforward. The word usage and definitions correspond to the *Industrial Engineering Terminology*[2] wherever possible.

For convenience, the text is written as if the reader were familiar with all three computer programs, even though this is not actually necessary. It also assumes that you will be using the diskette. You can construct the files without using the diskette, but you will waste a lot of time entering the data into your computer if you do not use it.

The cost of the diskette will soon appear to be insignificant compared to the time lost in this manual function—time that could be spent designing your layout. The diskette is designed to assist you in completing the plant layout project as quickly as possible.

Simplicity and speed are fundamental to the techniques described in this book. See Fig. 1.1, which is an example of how your layout will appear in two dimensions (2-D). The equipment templates are simple, but since they are numbered, they identify each piece. Figure 1.2 shows how your layout will appear in three dimensions (3-D). Again, the equipment templates and figures of people are simple, yet the appearance of the work area is adequately portrayed and can be evaluated.

The spreadsheets are not formatted for a formal presentation, yet the information can be assimilated quickly. The checklist items are brief, yet very informative. Read this book carefully. The information is presented in a concise form, and would normally be contained in a book twice as large.

For the above reasons, the writing style is not highly technical, and the instructions can be absorbed on the fly. Much of the

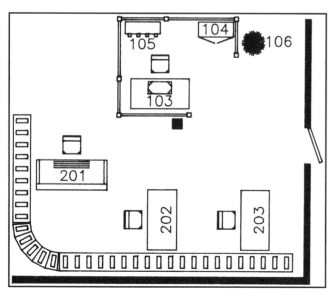

Figure 1.1   2-D layout demonstration.

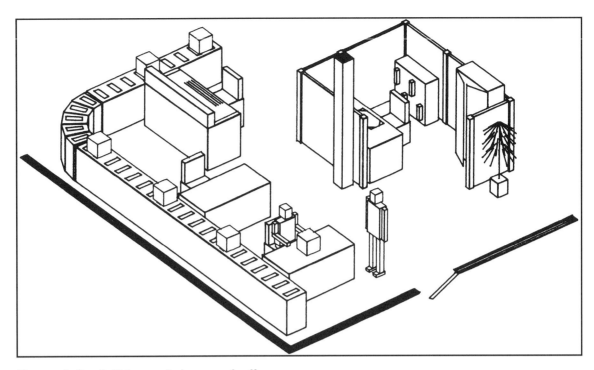

Figure 1.2   3-D layout demonstration.

information is presented in precise detail for the first-timer but can be passed over rather quickly by an experienced designer. You will not have to pause to consider ambiguous word combinations or remember unfamiliar acronyms. Most of the acronyms and abbreviations used in this book are common to the industry:

IE   *Industrial engineer.* The term *IE* is used here to describe the industrial engineer who is the plant layout project leader for the example company, the Flexible Division.

QA   *Quality Assurance.* The inspection group that has the final responsibility for product quality.

ESD   *Electrostatic discharge.* The static charge that accumulates on all equipment and operators and that can discharge to ground through electronic devices, weakening or destroying them.

JIT   *Just-in-time.* A system of manufacturing that accentuates the elimination of waste in all forms.

WIP   *Work in process.* A term applied to the buffer inventory between work stations. JIT implies no WIP.

MRB   *Material Review Board.* The group of people who are responsible for the disposition of vendors' parts that have been rejected by the production area.

EMF   *Electromagnetic field.* The magnetic field surrounding every conductor that is electrically powered. It is believed to be hazardous under certain conditions.

ADA   *Americans with Disabilities Act.* This act became law in 1990, and compliance under Title III is required for all employers.

OSHA	*Occupational Safety and Health Act.* A federal law that controls certain physical conditions that every employer must adhere to regarding the health and safety of the employees.
HVAC	*Heating, ventilating, and air conditioning.* The entire air and temperature control system of a building.
CEDAR	*Cost Evaluation Data And Report.* A product cost analysis spreadsheet, developed by the author.

This book promotes a philosophy of dealing with problems that has always been obvious but is so widely disregarded that it must be repeated and clarified. That philosophy is simple: Solve the problem, don't treat the symptoms. Locating the problem and solving it can be a long and difficult task with little external reward, while treating the symptoms is relatively easy and provides immediate recognition. For that reason, treating symptoms of problems has become pervasive in much of the world, in all levels of government as well as in business.

Here is an example of a problem and how it is commonly treated in the manufacturing business:

1. The *symptom:* Units are being rejected by some production process equipment. The common response is to treat the symptoms by constantly adjusting the equipment so it produces a higher percentage of good parts.

2. The *problem:* A variable quality level of incoming material for that process is passing through incoming inspection because of loose inspection parameters.

3. The *cause:* The vendor is not supplying material at a sufficiently high quality level. Notice that the inspectors, using faulty incoming inspection criteria, are not the real cause of the problem.

4. The *solution:* Work with the vendor to raise the material quality or find a new vendor.

Here is an example of a problem and how it is commonly treated by the federal government:

1. The *symptom:* Homeless people are roaming the streets. The common response is to build shelters for them.

2. The *problem:* These people cannot afford adequate housing.

3. The *cause:* Many of these people are not employable because they lack skills and education.

4. The *solution:* In the short term, investigate how to educate and motivate each subgroup so members can earn a living. In the long term, provide a protected environment that is tailored for each subgroup of children so they will be employable as adults. As you can see, the solution is extremely arduous with no external reward, so it will never be achieved.

The worst part about treating symptoms rather than solving problems is that the problem continues forever and the symptomatic treatment dogs its trail. When you are faced with what appears to be a problem in manufacturing, recognize what it really is: a symptom of something much larger. Trace the difficulty upstream, and do not stop until you find the cause. Then identify the parts of the difficulty: symptom, problem, and cause. Only then will you be able to figure out the true solution. If the symptoms appear minor compared to the problem, you have probably found the real problem.

The CAD Program

There are so many advantages to designing your plant layout on a CAD system that the manual drafting method appears archaic. As you will see, many different plots of the same file can be used to portray different views and versions of the plant layout drawing, and they will all be updated together. For instance, a mechanical renovation drawing can be plotted quite easily after turning on one layer and turning off others. And, since all layers are in the same file, they will all be updated automatically as the equipment and walls are relocated during the revision process.

The AutoCAD program has the advantage of being very popular in the manufacturing industry. As a result, once you learn to use it, your knowledge will be valuable to almost any employer. If your company already has AutoCAD installed on your computer network, you do not have to enroll in a formal instruction class to get started. Just study the manual and start experimenting.

A host of commands form the complete program, but only a few are required to design a layout and use the diskette files efficiently. Here is a list of the commands required to make a plant

Uncomplicated		Complex
BLOCK	LTSCALE	BREAK
CIRCLE	MOVE	CHANGE
COPY	OOPS	DIM
DIST	ORTHO	ELEV
ERASE	PAN	HATCH
FILLET	POINT	LAYER
GRID	PURGE	PLOT
HIDE	QUIT	TEXT
INSERT	REDRAW	TRACE
LIMITS	REGEN	VPOINT
LINE	RENAME	ZOOM
LINETYPE	SAVE	
LIST	SNAP	

layout using this book, the diskette, and AutoCAD. The commands in the first two columns can be learned quickly and require only one or two keystrokes to execute. The commands in the third column will require more time to learn but still are not difficult.

Once you have started using a CAD program, a vast array of possibilities will open up that formerly seemed too time-consuming to tackle. A complicated flowchart, for instance, can be approximated on the first draft and then revised as often as necessary until it represents the process exactly. So many individual drawings would be required to achieve this manually that the project would never get off the ground.

It is possible, of course, to design a plant layout with a less universal CAD program, but then, as your talents expand, the program may not expand with you. Also, a vast amount of third-party software is available for AutoCAD that can satisfy your every need regardless of your field of experience.

All the layout drawings in this book are full-scale replicas of actual plots. You can measure the equipment and the space around the equipment right from the illustration if you want to determine exactly what dimensions were used. However, do not measure your *copies* of these illustrations. They may not be to scale since copy machines often reduce the image slightly.

The Software

The software described in this book can be purchased from the supplier shown on the last page. The diskette is available in $5\frac{1}{4}$-in (1.2-Mbyte) or $3\frac{1}{2}$-in (1.44-Mbyte) formats. The AutoCAD files on the diskette can be used on MS-DOS or Windows computers listed in *The AutoCAD Resource Guide.*[3]

If you are using programs different from those mentioned in this book, the files may still be compatible. Most new word processing and spreadsheet programs can translate these files into their own formats quite easily. Some CAD programs can even convert an AutoCAD file directly to their formats.

AutoCAD LT[21] is a less expensive, scaled-down version of AutoCAD. The AutoCAD .DWG files on the diskette can be used directly by AutoCAD LT. If you plan to use AutoCAD LT, ask Autodesk about its latest 3-D capabilities. Find out if you can change the elevation and thickness of traces, and if they will plot as 3-D solid models.

All AutoCAD files on the diskette have also been converted to a DXF format, so virtually any CAD program that will accept this format can use these files. Consult your program manual for this capability before ordering the diskette.

Caution must be used in converting AutoCAD or DXF files to your program. This conversion process is not always perfect, since there are a wide variety of CAD program types. For instance, if the text fonts do not match, the process may abort and refuse to complete. For that reason, all the text in these files has been converted to the STANDARD TXT font.

The term *3-D* is used in this book to indicate a three-dimensional solid model that can be viewed from any direction. It is not a wire-frame model. The initial computer view of the object may be in the wire-frame form, but the hidden lines can be removed with a HIDE command or its equivalent.

There may also be problems with converting AutoCAD traces. All the 3-D blocks in the AutoCAD files are composed of 3-D traces which may convert to 2-D lines in your program, even if it is capable of a 3-D presentation.

Problems can occur when sub-blocks nested within blocks are converted by some programs using the DXF file. These sub-blocks are sometimes removed from the block during the conversion process and are assigned a different elevation from the original block. For that reason, all blocks on the diskette are composed of primitive entities and should remain together as a unit during conversion.

If you use a program that is compatible with AutoCAD, the conversion can still have problems. For instance, Generic CADD (Release 6.0)[4] is a less expensive 2-D program that can convert an AutoCAD file directly, but does not have a trace command. In the direct conversion, the traces are changed to filled areas, which disappear when fill is turned off. This is totally unacceptable.

However, when Generic CADD converts the DXF file, the traces appear as perfectly acceptable 2-D lines. The blocks are changed to components, the file extension is changed from .DWG to .GCD, and the program is ready to go.

Another program, CADKEY,[9] will not convert AutoCAD files directly but is still able to convert the DXF files successfully. The AutoCAD blocks are converted to 3-D models and can be plotted to produce illustrations that are similar to those in this book.

In all cases, if you plan to use the diskette and convert the files to a CAD program other than AutoCAD, first check in your program manual to be sure that the program will accept the DXF format with the DXFIN command or equivalent. Then determine from the manufacturer how your program will convert an AutoCAD 3-D trace and how it will appear on your screen.

All the files on the diskette were created by using these programs: AutoCAD (Release 10), Lotus 1-2-3 (Release 2.2), and WordPerfect (Version 5.0). The files will be compatible with any of these programs with the same revision or higher. Each type of diskette contains all the files listed in Fig. 1.3. Complete instructions for using each file are given in this book.

The Model Factory

The Flexible Division is a fictitious company that performs a layout project right along with you at every step. If there is an existing company of the same name, no reference to that company's

The files described here are on a diskette that can be purchased from the supplier shown on the last page of the book. The diskette is available in 5¼-in (1.2-Mbyte) or 3½-in (1.44-Mbyte) formats for MS-DOS and Windows computers.

AutoCAD (Release 10) and DXF files:

LAYOUT.DWG and LAYOUT.DXF. This prototype plant layout file can be used to start your plant layout drawing. Various views of this drawing are shown in

Figure 1.1   2-D layout demonstration.
Figure 1.2   3-D layout demonstration.
Figure 7.4   Blank layout drawing.

FLOW.DWG and FLOW.DXF. This Flexible Division old process flowchart can be used to start your flowchart drawing. The initial view of this file is shown in

Figure 2.2   Old process flowchart.

PIE.DWG and PIE.DXF. This set of Flexible Division pie charts can be used to start your pie chart drawings after you have made a flow time analysis. The initial view is shown in

Figure 4.3   Flow time pie charts.

RELATE.DWG and RELATE.DXF. This Flexible Division relationship diagram can be used to start your relationship drawing. The initial view is shown in

Figure 5.2   Relationship diagram.

LIBRARY.DWG and LIBRARY.DXF. This block file contains the entire library of 3-D blocks and can be inserted into any drawing. The initial view of all blocks is shown in

Figure 8.2   Block insertion file.

LEGEND.DWG and LEGEND.DXF. This block file contains 2-D examples of the common equipment blocks in the library. It can be inserted into any layout drawing to provide a legend for the blocks. The initial view is shown in

Figure 13.3   Layout legend.

LOTUS 1-2-3 (Release 2.2) files:

LABOR.WK1. This Flexible Division spreadsheet file can be used to start your old and new labor analyses. The initial display is shown in

Figure 4.1   Old CEDAR Labor Analysis.
Figure 4.2   New CEDAR Labor Analysis.

Figure 1.3   List of files on diskette.

LIST.WK1. This Flexible Division spreadsheet file can be used to start the list of all the equipment in your new plant layout and to determine the approximate floor space required. The initial display is shown in

Figure 5.1   New plant equipment list.

WordPerfect (Version 5.0) files:

BLOCKS.WP. This list of filenames of the blocks in the library can be used as a reference, and you can add the filenames of your custom blocks to it. The initial display is shown in

Figure 7.1   List of blocks.

MOVEPLAN.WP. This Flexible Division move schedule can be used as a starting point to plan your move and construct your move schedule. The initial display is shown in

Figure 15.1   Move schedule.

Menu file for AutoCAD:

MACROS.MNU. This menu can be loaded into the LAYOUT.DWG file or any AutoCAD file that contains the right layer names. The 50 macros that are described in the book will then appear on two screen menus. The macros are described in

Figure 9.2   Macro menu explanation.

DOS text file:

MACROS.DOS. This DOS text file contains the same 50 macros that are in the MACROS.MNU file and can be loaded into an AutoCAD menu. Instructions for their use are provided in the book.

Quick Schedule Plus file:

SCHEDULE.QS. This Flexible Division plant layout project schedule can be used only with the Quick Schedule Plus program. The initial display is shown in

Figure 6.1   Plant layout project schedule.

Figure 1.3   List of files on diskette (_Continued_).

operation is implied. The IE, who is the plant layout project leader for the Flexible Division, is faced with problems similar to yours, develops original solutions, and serves as your guide throughout the project. The IE makes the following items for the Flexible Division so you can see what is actually required:

1. A process flowchart
2. A labor analysis
3. Two pie charts
4. A new plant equipment list

5. A relationship diagram

6. A plant layout project schedule

7. Many custom equipment blocks

8. A new plant drawing

9. An old plant layout

10. A new plant layout

11. Three utility overlays

12. A moving schedule

Their plant layout does not include a complete office layout since that function is performed in the home office building next door. A small office arrangement example is provided, however. Because the office layout is not needed, the example plant layout can be shown at a full $\frac{1}{8}$ in = 1 ft scale, and the manufacturing department can be better emphasized. The result is a layout drawing of a plant that is only 4100 sq ft in area, but contains all the essential elements of a manufacturing system.

The Flexible Division anticipated its move by more than a year. Thus, it was able to streamline its operation extensively and incorporate the improvements to the production system into the new plant layout. The Flexible Division planned for major improvements that would lower the product cost, improve the product quality, and raise the production rate per shift. Improvements of this magnitude can easily take more than a year to implement.

The Flexible Division plant is relatively small, but this book can be used to lay out a plant of any size, for any product, from a transistor development lab to an automobile production line. The capacity of your computer to handle the data is the only limit to the magnitude of the project.

You will see how the Flexible Division:

1. Improves the smoothness of its product flow.

2. Drastically reduces the flow time of the product.

3. Incorporates a JIT production plan.

4. Eliminates the WIP between assembly stations.

5. Reduces the stores' inventory by careful planning and vendor qualification.

2

Process Flow Analysis

Process flow will be discussed in this chapter. A definition of the term *flow* will be presented that may or may not be new to you. Different categories of flow and methods for measuring flow will also be given. The flowcharts used by the IE to analyze the Flexible Division material flow will be illustrated.

Introduction

A company can realize the maximum benefit from a plant layout only if it is the culmination of a true production improvement project. Then the layout will not only incorporate the obvious improvements, but also lead to a new method of manufacturing that is much more efficient and has a lower product cost. This pre-layout improvement process might take months, even years, but it can have a profound influence on the entire production process.

Of course, in the real world, most plant layouts are performed at the last possible moment because of time restrictions placed on the designer. If that is your situation, you will be forced to skip to Chapter 5 and start with the immediate concerns of getting the project under way.

The terms used in this book to describe the various stages of assembly are as follows:

1. *Product* describes the finished assembly as it is ready to be packaged and shipped to the customer.

2. *Unit* describes the main chassis or assembly as it progresses through the assembly process until it becomes the product.

3. *Part* describes any piece of material that will be assembled into the unit.

Definition of Flow

A common thought expressed by management is, "Let's make a new layout and improve the flow." The production personnel in the audience have an intuitive sense of what the speaker means, and they know why the "flow" needs to be improved. Even if the product appears to be traveling smoothly through the production process, everyone usually agrees that the flow can be improved. When the manager says "flow", this conjures up an image of everything that is concerned with efficient production. This description is still true, even under the expanded definition of flow presented in this book.

Flow analysis is an art, not a science. Because of human involvement, there are no conditions that lead consistently to the same result. Plant layout, which is influenced heavily by flow analysis, is very much an art, as demanding and frustrating as trying to paint your first landscape.

If the term *flow*, in manufacturing, refers to a condition of manufacturing smoothness, then how can it be measured? Surprisingly, it can be measured by an element of time—the time it takes to complete a segment of the production process, referred to as the flow time. The *flow time* is the average amount of time that it takes for one unit to pass through a segment of the production system.

For instance, the flow time for a segment of production might be one hour, starting when one unit is placed in position, ready for the first operation, and ending when that same unit is placed in position for the next operation. If there are conditional pauses in the sequence, such as rework loops, then that time must be prorated and added to the flow time. The flow time then becomes a measure of the flow, or the overall efficiency, of the segment under study.

Flow is one of those relative words that has little utility unless it can be quantified. Without a time measurement, what is improved flow? Without a time measurement, improved flow can only be evaluated by the judgment of the beholder. Flow time is the quantitative measurement of flow. To determine if the flow has been improved for any segment of the production process, the flow times, before and after the change has been made, must be compared.

As an example, suppose you have an assembly line that is almost chaotic, and you decide to reorganize the assembly sequence into a smoother flow. First, measure the total flow time for one unit to pass through the assembly process. If there are rework cycles, multiply the average rework time by the average percentage of units that are reworked, and add the result to the flow time. Record all the time periods that the unit spends waiting for the next operation. Monitor the unit closely as it moves through the assembly process to ensure that every time element is recorded and that it represents an average unit.

Then reorganize the layout, change the process sequence, retrain the personnel, and make any other improvements that can be imagined. After the new process has stabilized, measure the new flow time for one individual unit. The new system may appear to be smoother, but if the two flow times are equal, you have not improved the flow. Of course, other improvements may have been realized, such as in unit quality or capacity, but the flow is still the same.

Flow can be imagined as a river, with a bottle floating on the water. The time it takes for the bottle to get from the bridge to the waterfall is the flow time. It also includes the average time that the bottle spent in the whirlpool by the old mill.

Significance of Flow

If improving the production process is one of the goals of the plant layout project, then flow analysis is the first step toward that goal. It is the flow that characterizes all the principles of manufacturing efficiency. It is the flow that encompasses most of the principles of the JIT manufacturing philosophy. It is the flow that can be quantified, justified, and monitored by the project manager. The flow is the most important aspect of the production process that will affect your layout. Manufacturing process flow can be divided into four categories:

1. *Storeroom flow.* The flow time required to receive, store, and deliver a group of parts to the production floor. The flow time clock starts when an order quantity of parts is delivered to the receiving dock. It stops when the last parts from that group are delivered to the production floor.

2. *Order flow.* The flow time required to convert a sales order to a work order for production. The flow time clock starts when a salesperson receives an order for a unit from a customer. It stops when the production work order is received by the storeroom and the production supervisor.

3. *Manufacturing flow.* The flow time required to manufacture a unit, including the parts preparation in the storeroom. The clock starts when the production order is received in the storeroom. It stops when the completed unit is delivered to the finished-goods inventory area.

4. *Shipping flow.* The flow time required to receive the unit into finished-goods inventory, process, pack, and ship the unit. The clock starts when the unit is received into finished goods. It stops when the unit leaves the shipping dock.

Category 1 is a measure of the efficiency of the storeroom. Category 3 is often called the *throughput time.* The total time of categories 2, 3, and 4 is referred to as the *lead time.* That is, the lead time is the time required by a supplier to fill an order, from the time the order is received until the product is shipped to the customer. All the categories shown above can be analyzed, investigated, and shortened to produce a more efficient operation.

Why should it take three days to convert a sales order to a production work order when the actual work involved is less than 30 minutes? Every minute a production order waits in someone's in-basket, the delay is costing the company money in the form of profit. Every minute a part is idle on the receiving dock or in the storeroom or on the production floor, the lost time is costing the company money in the form of profit. The elimination of wasted time is a constant, ongoing endeavor that must be diligently pursued. The success of your company depends on it.

Flow and JIT go hand in hand as tools for improving the manufacturing process. JIT is a broader concept which cannot be quantified as easily as flow. On the other hand, the JIT concept of quick delivery and small order quantity is a major source of improving the flow.

The significance of improving the flow is derived from the cost of time—not the labor time, but the material idle time that is hidden in every operation. If the salespeople from your company can quote a more competitive shipping date because you have a shorter flow time, that factor will attract customers like a half-price sale.

This is especially true as more customers and vendors become aware of their own flow, and demand immediate deliveries. In the future, it is conceivable that all flow-conscious manufacturers will be banded together in a common goal, and all others will be on the road to anonymity. Most Japanese manufacturers have formed such a bond already.

In the manufacturing and off-line support group areas, the idle time for parts is a double-barreled blast at profits. One shot hits the profits because of long lead times and lost sales. The other shot hits the profits because the investment interest must be paid on all parts as soon as they enter the receiving door. When units or parts are standing idle, the wasted investment interest starts accumulating. That includes items sitting on the shelf in the storeroom as well as waiting around overnight for the day shift to start.

The importance of time and flow is not a new concept. Like every basic principle of modern manufacturing, it was pioneered by Henry Ford in the early 20th century. In 1926, at the Rouge River plant near Dearborn, Michigan, Ford set in place a monument to manufacturing flow that has never been equaled in modern-day production.

Ford started timing his operation when the iron ore barges arrived at the Rouge River plant from Minnesota and stopped timing when the Model T containing the steel from that ore rolled off the assembly line. At the height of production created by his genius, that total time was 28 hours. They didn't call it JIT in those days. To use his methods was called "to Fordize," and it meant the constant movement of material from the time it was mined until it was delivered to the customer.

Flow improvement was just one of Henry Ford's contributions to the manufacturing industry. Others are the moving assembly line, vendor quality control, self-inspection by operators, hiring the handicapped, minimum middle management, vertical integration, orderliness of the shop floor, elevated pay scales for productive workers, elimination of waste in all forms, and many more. More information is available in a biography of Henry Ford entitled *Henry Ford: The Wayward Capitalist*.[18]

The manufacturing department of your plant is normally responsible for the largest category of flow time. As project leader of the new manufacturing area layout, you are in a prime position to analyze and improve the flow.

Conventional Process Flowchart

The first step in analyzing the flow for an area is to specify the tasks that make up the flow in the order in which they are performed. A conventional process flowchart does this clearly, but does little else. Figure 2.1 is a good example. It adequately describes the flow of operations for one product of the Flexible Division. It might have some value as a presentation to management or as a wall chart to guide visitors on a tour of the manufacturing floor, but as an analytical tool, it lacks detail.

Figure 2.1 describes the flow through the manufacturing area of the most expensive RAM/ROM Memory Bank (Model 107) manufactured by the Flexible Division. The product case is a computer-gray module about 12 in wide by 12 in deep by 8 in high. It is composed of a control card and a series of solid-state flash memory cards that plug into a back plane card secured to the chassis. A microprocessor, a numeric keyboard, and an LED display are used to program the memory status of the bank.

The resistors and capacitors of one thick-film microcircuit are trimmed in a unique pattern for each customer to provide a code that denies unauthorized access. The memory bank can be programmed, wholly or partly, as a random access or a read-only database, and it does not require battery or line power to retain the memory. The maximum capacity of this product is 7 gigabytes (Gbytes) of nonvolatile memory, plus 3 Gbytes of redundant arrays.

The example flowchart shown in Fig. 2.1 was designed by the IE, who had just been hired to head up the plant layout project at the Flexible Division. At this stage of flowchart development, the IE tabulated the operations, but did not measure the flow time. These are the assembly operations. The microcircuit part, an off-line assembly, is shown last:

1. Assemble the chassis.
2. Assemble and mount the back plane card.
3. Build the card cable.
4. Build the main cable.
5. Connect the cables.

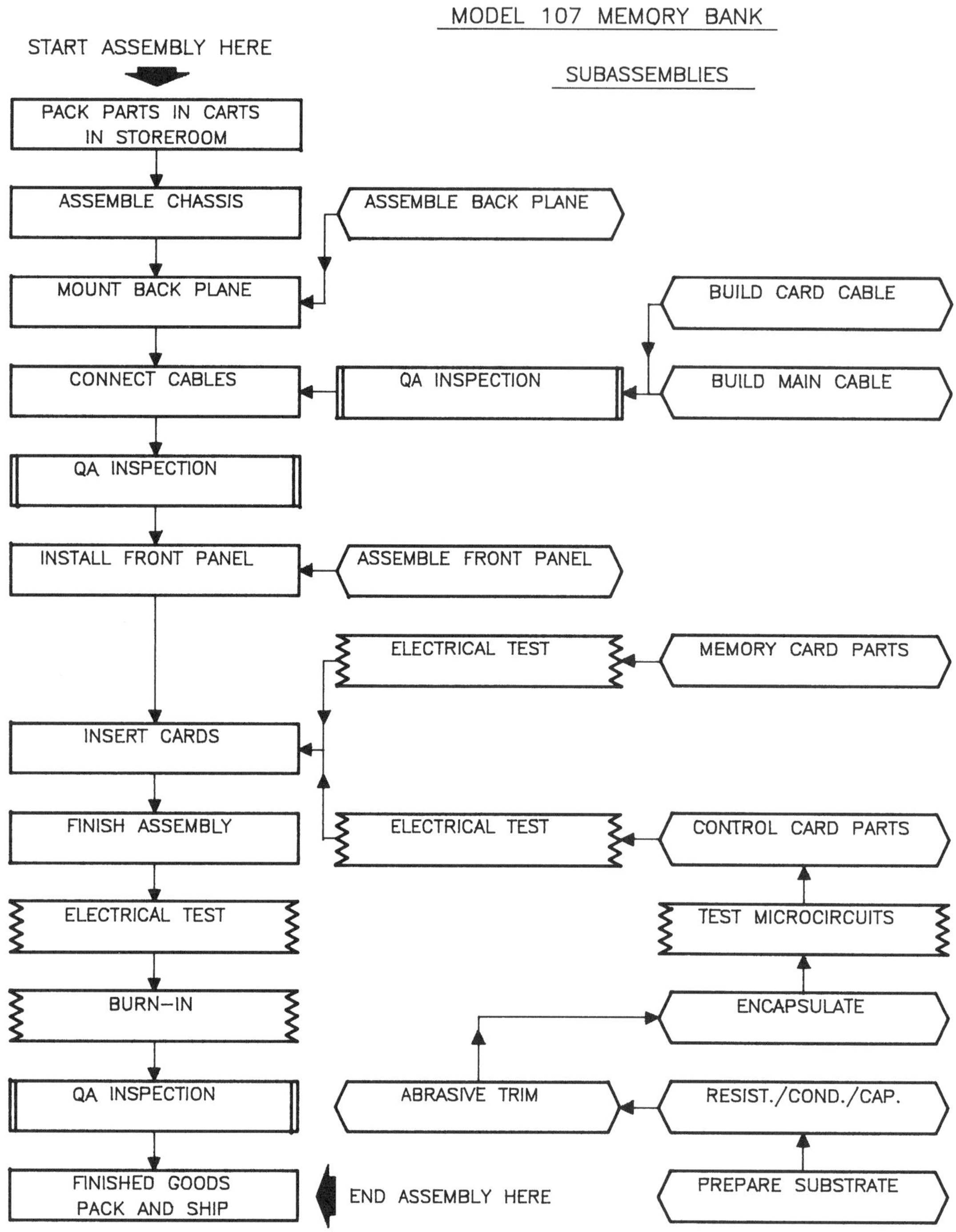

SUBASSEMBLIES

START ASSEMBLY HERE

```
PACK PARTS IN CARTS
IN STOREROOM
        │
        ▼
ASSEMBLE CHASSIS          ASSEMBLE BACK PLANE
        │                         │
        ▼                         ▼
MOUNT BACK PLANE ◄───────────────┘
        │                                    BUILD CARD CABLE
        ▼                                           │
CONNECT CABLES ◄──── QA INSPECTION ◄──── BUILD MAIN CABLE
        │
        ▼
QA INSPECTION
        │
        ▼
INSTALL FRONT PANEL ◄──── ASSEMBLE FRONT PANEL
        │
        │         ELECTRICAL TEST ◄──── MEMORY CARD PARTS
        │                │
        ▼                ▼
INSERT CARDS ◄──────────┘
        │                ▲
        ▼                │
FINISH ASSEMBLY   ELECTRICAL TEST ◄──── CONTROL CARD PARTS
        │                                       ▲
        ▼                                       │
ELECTRICAL TEST                         TEST MICROCIRCUITS
        │                                       ▲
        ▼                                       │
BURN-IN                                 ENCAPSULATE
        │                                  ▲
        ▼                                  │
QA INSPECTION      ABRASIVE TRIM ◄──── RESIST./COND./CAP.
        │                                  ▲
        ▼                                  │
FINISHED GOODS     END ASSEMBLY HERE  PREPARE SUBSTRATE
PACK AND SHIP
```

Figure 2.1   Conventional process flowchart.

6. Install the front panel.
7. Assemble and test the control and memory cards.
8. Insert the cards.
9. Finish the assembly.
10. Electrical test the finished unit.
11. Burn in the finished unit.
12. Pack and ship the product.
13. Microcircuit assembly:
    a. Prepare the substrate.
    b. Screen and fire the conductor, resistor, and capacitor layers.
    c. Trim the components to value.
    d. Screen and cure the epoxy encapsulant.
    e. Electrical test the finished microcircuits.

Old Process Flowchart

Although the flowchart shown in Fig. 2.1 describes the operation sequence, it does not provide the additional information needed to improve the flow. The next step taken by the IE was to modify the flowchart and use it as a tool to improve the production system.

The manufacturing department of the Flexible Division had decided to start its improvement program more than a year before the scheduled move into the new building. The IE was assigned the task of improving the flow and making a new lay-out that would incorporate those improvements. To get started, the IE examined the existing production process for ways to improve the process flow. The IE identified some conditions that restricted the flow and developed a method of including them on the flowchart.

The IE then modified Fig. 2.1 to include and emphasize these restrictions. The revised flowchart, Fig. 2.2, describes the process flow of the Flexible Division before any improvements were made, and illustrates many key items that restricted the flow. The additions that the IE made to the conventional process flowchart are listed here. They highlight many of the problems that must be solved to improve the flow in almost any manufacturing facility and will be explained in more detail in later chapters.

1. *Parts (boxes on both sides of the flowchart).* Most of the parts that are assembled in each operation are labeled and shown in these boxes on the chart. By including the flow of parts, the flowchart gives a more complete picture of the actual assembly process. Also, more expensive parts can be recognized and tracked.

2. *Queue dots (dots in the lower right corner of each operation box).* If a dot is shown, it means that the units are stored as work-in-process (WIP) after that operation. These queue dots are the most important indicator of lost time on the assembly line. For the Flexible Division, they indicate that units are stored on a rolling cart between each operation. After inserting the queue

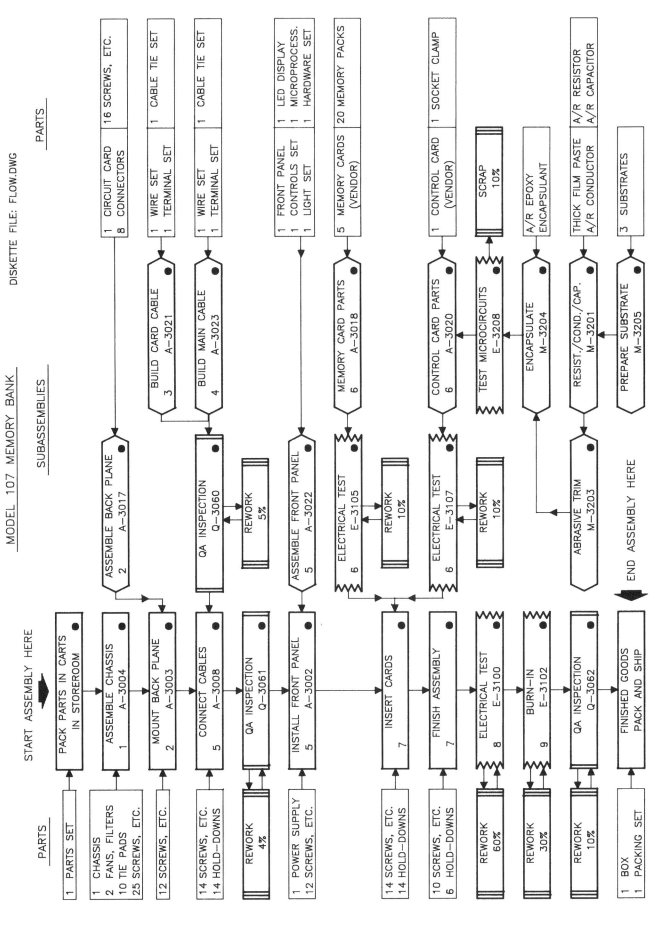

DISKETTE FILE: FLOW.DWG

Figure 2.2 Old process flowchart.

dots on the flowchart, the IE immediately developed a way to eliminate them.

3. *QA rework loops (attached to the QA inspection boxes with double arrows).* When the QA inspector rejects a unit, it must be returned to the operator who caused the reject, or returned to a common area where repairs are made. These QA inspections and reject loops add nothing to the value of the product, and are prime targets for flow improvement. The waste in time and labor is obvious, but the solution can be realized only by a dedicated effort from everyone involved.

4. *Electrical test rework loops (attached to the electrical test and burn-in boxes with double arrows).* When the electrical test technician rejects a unit, it must be repaired on the spot or returned to a rework bench. The electrical test and the rework loop add nothing to the value of the product either.

5. *Percent scrap (next to the test microcircuits box).* Some rejected parts cannot be repaired, or the repair effort is not cost-justified. In this example, the thick-film microcircuit is a nonrepairable part, and the reject parts are scrapped. The improvement of this condition is an ongoing process control effort that is constantly monitored by the Flexible Division.

6. *Percent rework (percentage shown in the rework boxes).* This figure indicates what percentage of the units is rejected. All the units will be reworked. The percentage of units involved is very important when you are measuring the flow time. If this percentage can be reduced, not only are the flow time and the unit quality improved, but the rework labor is reduced as well.

7. *Operation numbers (lower left corner of each operation box).* They assist the layout designer in identifying the operation performed at each work station, and they will also be shown on the layout.

8. *Process instruction numbers (numbers shown directly below each operation in the box).* When the IE entered the numbers of the approved instructions on the flowchart, the missing ones became immediately obvious, operation 7, for instance. The operation instructions are crucial to a streamlined and efficient production system, and they must be written and approved for all activities before a uniform product can be produced. These instructions can greatly affect the flow and product quality when new operators are being trained, when assembly procedures are updated, or when there is a question about how an operation should be performed.

When the IE started the flow improvement project, the production system was essentially a job shop. It was not very efficient, and assembly errors were common. The company's business had grown rapidly because its product was unique and demand was very strong. Its production plan developed haphazardly because the time and personnel were not available to design a smooth system. As a result, the system was not planned, but expanded instead into a disconnected frenzy of assembly activity. The situation was ideal for streamlining, and making a new plant layout would implement the new plan.

You may not have that much time to concentrate on improvements in your plant before starting the plant layout. Although the move to a new plant may be anticipated by a year or more, the plant layout project will typically not start until the last possible moment. As a result, a valuable opportunity to improve the production process is lost, an opportunity that may never come again.

The management of your company may be planning a plant relocation some time in the future and deciding what steps will be needed to complete such a project. That is your opportunity to seize the moment and propose a plan to perform a major reorganization of the production system, from receiving raw parts to shipping the final product.

If that opportunity is at hand, the first step in improving the manufacturing process is to make a process flowchart similar to Fig. 2.2. By doing so, you will become familiar with each step of the assembly, and the finished chart will flag those items that contain the greatest potential for improvement. Making this chart will involve numerous interviews with the direct labor people and their supervisors.

While you are constructing the flowchart, make an exhaustive investigation. Each interview will add more information, and will allow you to fine-tune the flowchart to perfection. That is important—make the chart as accurate and informative as possible. At this stage, you can accept the operator's estimate of rework percentages; then later you can gather data if more accuracy is needed. Do not be satisfied with the chart until everyone agrees that the data you have illustrated is a true picture of the assembly process.

Many computer programs are available for constructing flowcharts, but none can match the versatility of a good CAD program. As you develop more flowchart attributes and symbols that reveal problems unique to your industry, the CAD program will be able to illustrate them easily.

New Process Flowchart

The benefit of a new plant layout can be far greater than just facilitating a move to a new building. The new layout can start your company into a new mode of operation, create higher profits, and result in a product of superior quality.

The start-up of your new assembly operation is like opening night at a Broadway play. Your new plant layout sets the stage, and the curtain opens as the first unit is assembled. The production manager is the producer. You are the director and the scriptwriter—the script is the process flowchart. If you use the same old script that you used for the last play, only a few people on break will watch.

However, if you redesign the script for a new, low-cost, and streamlined assembly system, there will be standing room only in the audience, and your boss will be sitting front row center. It's your choice.

The Flexible Division IE reviewed the flowchart shown in Fig. 2.2 to find opportunities for improvement, and found so many that several studies would be needed to research their practicality. To get started, the IE developed the new flowchart shown in Fig. 2.3. Notice that many operations were completely eliminated. At this stage, the IE only suspected that these operations were superfluous. If they could not be eliminated, they would have to be added back later.

This new flowchart was more than just a description of the new process; it initiated the entire reorganization. As the investigation proceeded, the IE made even more changes. All these changes defined the goals of the flow improvement project. They will be explained in more detail in later chapters and are briefly described here:

1. QA inspections—eliminated all three inspection operations.

2. Electrical test—eliminated three test operations.

3. Rework—eliminated six rework operations.

4. Burn-in rework—lowered the reject rate from 30 to 10 percent.

5. Burn-in cycle—reduced the burn-in cycle from 72 to 60 hours.

6. Queue dots—eliminated 18 cases of WIP.

7. Kitting of parts—organized the parts for each product into kits that would be used for each assembly operation.

All these improvements were originated more than a year before the move to the new plant was scheduled to take place. However, considering the magnitude of these changes, a concerted effort was required by all departments involved to organize these changes and have them documented before the move. Several full-scale projects were initiated by the engineering, materials, purchasing, quality assurance, and production departments. These projects will be described in later chapters.

Figure 2.2 is included on the diskette as an AutoCAD file called FLOW.DWG. It is designed to be plotted on B-size (11 × 17 in) paper, then reduced to 75 percent to A-size (8½ × 11 in) paper on your copy machine. When you load this file into your AutoCAD program, notice the outline on layer 0. It is 13.3 × 10 in, and when that is reduced to 75 percent, it will equal 7½ × 10 in. The outline should be turned off prior to plotting. The boxes should be plotted with a wide pen and the text with a narrow pen. There are notes above the outline that describe the block numbers for the boxes.

You can easily move and copy these blocks, change the text, and design a chart that will exactly represent your product flow. The chart can be expanded up to E-size paper and filled with boxes if required. Be sure to include all information that pertains to your process. The more relevant data that you enter, the more you will be able to use the chart to improve the flow.

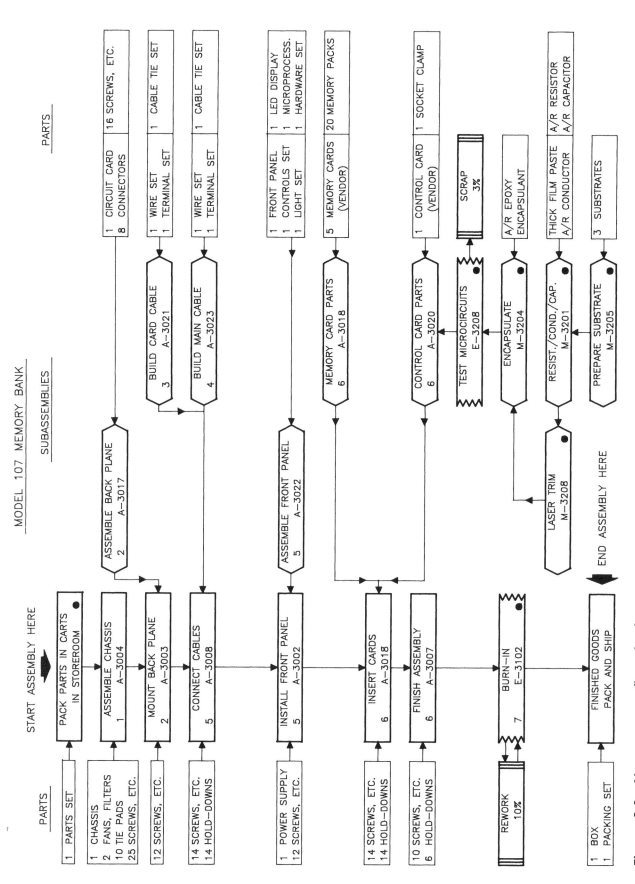

Figure 2.3 New process flowchart.

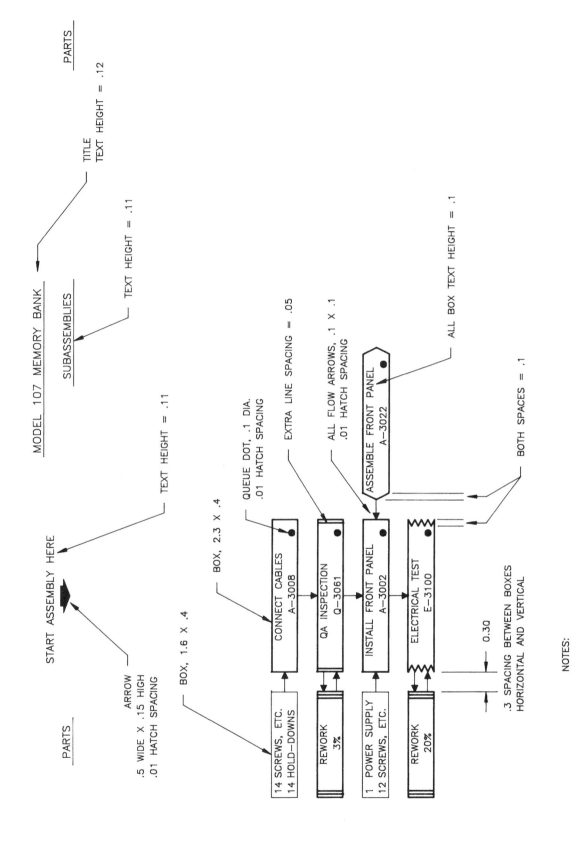

Figure 2.4 Flowchart dimensions.

The dimensions of the example flowcharts are shown in Fig. 2.4. If you use any other method of drawing your chart, you can use these dimensions to duplicate the example charts exactly. These charts were designed around a format that consists of an array of uniform-size symbol boxes. Notice the different box design for each activity, and that every box occupies a row and column location in the array. This type of layout enables you to display the maximum number of operation boxes on a single sheet in a well-organized format. By using flow lines with solid, well-defined arrows, the sequence of operations is also easy to follow.

3

Process Flow Improvement

Methods of improving the production flow of material will be discussed in this chapter. JIT will be discussed briefly, as it is beyond the scope of this book to provide detailed instructions for establishing a JIT program in your plant. Regarding your plant layout, however, flow improvement is important, and guidelines will be supplied in this chapter.

Flow and JIT

As stated before, a prime objective of making a new plant layout should be to improve the process flow, and the physical appearance of a production area will reflect that flow. A simple way to evaluate the status of the flow in your plant is to examine the conditions on the production floor. Any of the following items are a signal that the flow needs improvement:

1. Many units on carts, shelves, or conveyors waiting to be assembled.

2. Parts on the floor in bulk containers waiting to be assembled.

3. Shelving along the walls full of reject parts or other items that have not been disposed of.

4. Numerous rework benches, or a large amount of rework being performed on production benches.

5. Expensive machinery that is idle.

6. People expediting high-priority work orders.

7. Production status meetings being held every day.

8. Trash on the floor.

9. Anything in the aisles except people.

10. Operators making partial assemblies because of a part shortage.

11. Operators inspecting incoming parts to sort out the rejects.

In the following discussion, it is assumed that you are familiar with the basic principles of JIT. Process flow is a component of JIT, but JIT emphasizes the elimination of all waste in the production process. Process flow, on the other hand, emphasizes only the elimination of wasted time.

In an ultimate JIT system, any operator can shut down the manufacturing system if a unit is not immediately available to be assembled. That means the preceding operator is too slow, or the parts do not fit, or tools are missing, or other problems have occurred on the line. That is, almost any problem in the plant can ultimately affect an assembly operation and thus shut down the system until the problem is solved.

JIT has received an unfavorable reception by many companies because of this critical go/no-go situation. One well-known, high-tech company experimented with JIT by applying strict principles to one assembly line and then observing the results. During the first three months, the line was shut down 80 percent of the time because of problems that were uncovered by JIT. That situation improved very slowly over the next few months, but the test was finally abandoned as being too unwieldy. No modern, competitive manufacturer can long endure a system that allows only modest production while problems are being eliminated.

That is an example of the difficulty encountered when JIT is applied to the assembly line without preliminary planning: JIT will expose all the problems in the entire plant, including the quality of incoming parts. Finding the solutions to these problems becomes an endless long-term pursuit that can jeopardize the very viability of the company.

Many companies have abandoned their JIT programs for precisely this reason. That is the paradox of JIT: It was introduced as a means of streamlining the production system by uncovering production problems, but it is not being applied because it uncovers too many problems too quickly. A flow improvement project can face the same difficulties if it is initiated without careful planning.

If you want to start a flow improvement program in your plant, don't begin with the assembly area. If you do, the difficulties that you find will be symptoms only, caused by problems that are much further upstream. It is better to start with the cause of the problem and solve that first.

Start with a preliminary problem survey that will not interfere with the existing production system at all. Every person involved in the system, from purchasing to shipping, should participate in the program. They should all be given an opportunity to explain their problems and suggest solutions in a friendly, supportive atmosphere.

All participants should be interviewed one-on-one by the project leader, not in a group, not by their supervisors. If you try to organize a meeting for this purpose, a few people will dominate and most will not say a word. If the supervisor does the interview, the subordinate will hold back for fear of being labeled a troublemaker. Only you, as an unbiased audience, can provide the setting for all participants to vent their frustrations in a solo performance. When a participant presents an important contribution to you, be sure to give credit publicly where it is due.

After the survey is complete, develop a problem and solution report, and present it to management for discussion. The selected solutions can then be implemented in an organized, timely manner without interfering with production at all. It may take many months to solve some of these problems, but they are the very ones that will have the greatest impact on the system. This problem-and-solution program can improve the operation of your company dramatically. As an added benefit, it will instill in everyone a genuine feeling of participation in the company's success.

Categories of Flow Improvement

The categories of flow improvement shown here may not appear to relate to the flow, but each one will be described and validated later in this chapter. If time is not available for you to include flow improvements in your layout project, at least look over Fig. 3.1, which is a list of improvements. Some of them can be incorporated easily, such as aligning the equipment in the proper sequence instead of using a job-shop arrangement. The major methods of improving the flow are as follows:

1. Qualify the vendors who supply critical parts.

2. Reduce or eliminate independent inspections.

3. Reduce inventories of parts.

4. Alleviate critical bottlenecks.

5. Implement other flow improvements.

The term *critical part* is used in this book to describe a part that is routinely in short supply or is routinely rejected for any reason. The problems associated with critical parts can be responsible for many of the flow problems that affect the entire manufacturing system. These problems should be eliminated by a vendor qualification program before any other improvements are considered. Only those vendors who supply critical parts need to be qualified.

To start the program, locate the vendors who supply critical parts by questioning the buyers, supervisors, and inspectors. If the parts are never rejected, fine; but if they are rejected, then

This list of improvements will be used by the Flexible Division to improve its operation. Examine the list for items that relate to your plant, and then add more of your own. This list introduces many of the concepts advocated in the book; they will be explained in detail later.

1. *Vendors.* Critical-part vendors should be qualified. Noncritical-part vendors should be confirmed as reliable. In-line inspection stations can be eliminated on the layout.

2. *Receiving.* The shipping and receiving areas should be separated to avoid errors in routing. A low fence may be all that is required, and it should be indicated in the layout.

3. *Incoming inspection.* Incoming inspection should be a secure area with limited access. It should be used only to qualify critical-part vendors and spot-check reliable vendors. The flow of part containers should be controlled by a conveyor or equivalent apparatus, shown on the layout.

4. *Parts issue.* Parts can be issued in kits (color-coded tote boxes) for each subassembly, rather than the total quantities for each work order. Kit transporters should be shown on the layout.

5. *Storeroom.* The parts on the shelves can be arranged in kit order, rather than stock number order, to simplify the stock-pull operation. The floor area can be reduced when vendors are qualified or excess inventories are reduced, and the reduction can be reflected in the layout.

6. *Assembly.* The operations should be arranged in a logical sequence. Even a job shop will have similar products that dominate the production system. The flow pattern can then be based on these units. Flow arrows can show the results on the layout.

7. *QA inspection.* Quality inspections should be used only to monitor those operations that have not been qualified. QA stations should not be included in the direct flow pattern of the plant layout.

8. *Electrical test.* This test should be a brief monitor of the vendors' quality. If qualified vendors have their own environmental and test equipment, the electrical test operation should have a very high yield or should not be needed at all.

9. *Electrical test equipment.* Diagnostic test equipment should be produced for the vendor and the test station. A junior technician should be able to operate the station and, on rare occasions, locate a faulty part and replace it.

10. *Burn-in station.* Normally, without vendor qualification, the burn-in operation will suffer a 30 to 40 percent failure rate. If the vendors are performing their own burn-in test and troubleshooting, that rate will drop dramatically. Burn-in equipment and personnel requirements will be greatly reduced or eliminated. Those reductions can be reflected in the layout. This item applies to any critical test: electrical, mechanical, chemical, etc.

Figure 3.1   List of improvements.

11. *Finished-goods inventory.* This inventory is required to support the variation in product sales, and is an indicator of the flow status of the production area. If the flow is improved, the finished-goods inventory can be more accurately balanced against demand and the overall level can be reduced. As a result, the storage area can be made smaller in the layout.

12. *Thick-film process.* When any process is responsible for a low yield, that process must be studied in depth to determine the cause. In the case of the Flexible Division, the thick-film resistors were originally trimmed by an abrasion process that did not produce a very narrow-tolerance resistor. The new laser trimmer eliminated that problem.

13. *Conference rooms.* Conference rooms should be included in the layout only if needed for essential meetings, as defined in Chapter 9. They should not be included just for the purpose of holding production status meetings, which are a serious waste of time for key people. Status meetings are required only when the flow time for the products is excessive. Bottlenecks cause status meetings, not too many work orders.

14. *Supervision.* The production supervisor's desk should be in the immediate vicinity of the production area. This shortens the reaction time for problems on the floor and supports the unity of the group. It also reduces the need for a high-level line leader.

15. *Electrostatic discharge.* If this is a problem, conductive tiles or a conductive coating can be applied to the floor in the assembly area. An isolated earth-ground system can be provided for the operator's wrist strap or grounded smock. The system can be shown on your utility layout.

16. *110-V receptacles.* The receptacles on the workbenches should be modified to have ground fault circuit interrupters to provide a safer workplace. This item can be mentioned on your utility layout.

17. _____

Figure 3.1   List of improvements (*Continued*).

they are critical. There may be an interval of several months at this point while you conduct a failure analysis program to determine exactly which components are causing the problems. Gathering and evaluating failure data is an onerous task, but can be extremely rewarding when a key part is found to be the culprit.

Evaluate the need for any incoming inspection test, QA test, electrical test, or any test that is performed routinely on the units in production. All inspection and test operations add nothing to the value of the unit; they increase the flow time and are thus candidates for elimination. Some causes for rejection will be internal, not related to vendors. Other causes will be external, directly related to critical-part quality, and the vendor who supplies these parts should be qualified.

Make a list of the critical parts, the reason for their critical classification, and the vendor who supplies the parts. It will be your guide for initiating the vendor qualification program.

Vendor Qualification

Some companies have tried to implement a flow improvement program and reduce their stores' inventory of critical parts without first qualifying their vendors. The results can be disastrous:

1. A shortage of parts soon occurs either by rejection on the line or by late delivery.

2. The situation becomes serious as the operation shuts down.

3. The purchasing department becomes chaotic as it tries to expedite the resulting surge of "rush" orders.

4. The production personnel attempt to keep busy by making partial assemblies and stacking them in huge piles of work-in-process (WIP).

5. The vendor becomes irritated because this situation has never happened before even though good parts have always been in short supply.

6. In extreme cases, the reject parts may be assembled on a temporary waiver. Field failure rates go up, and the company's future becomes questionable.

Vendor qualification is the key to eliminating this situation, and it sets the stage for improving the flow. If flow is a monument to efficient production, then vendor qualification is the pedestal for that monument. Imagine how easy it would be to manufacture a product if every part or material received were perfect in every way. No testing would be required. If operators were well trained and the processes acceptable, no inspections would be required. Field failures would be almost nonexistent. If the product were reasonably priced, the manufacturer would completely dominate the industry.

Vendor qualification can have a dramatic effect on the long-term success of your plant layout and the future of your company. Rework stations, QA stations, and other test stations can be reduced or completely eliminated. The storeroom inventory and floor space can be drastically reduced.

A qualified vendor is much more than just a parts supplier. When a vendor is qualified to furnish parts to your plant, that vendor becomes virtually part of your operation. The term *4Q* is used in this book to describe a vendor who is qualified in every way to supply perfect parts. The four Q's that describe a qualified vendor stand for:

1. *Quality.* The vendor must demonstrate the ability to manufacture a part at a given quality level. That should include a program, initiated by the vendor, to qualify the vendor's vendors. The parts must be manufactured in a cost-effective manner, and the quality must be constantly refined.

Each lot of parts must be sample-inspected at the vendor's plant before it is shipped. This inspection must be at a somewhat higher quality level than was used by the incoming inspection department of your plant before qualification. If the vendor's sample fails, the vendor will screen the entire lot for rejects and will sample the lot again. Only qualified lots will be shipped.

2. *Quantity.* The vendor must demonstrate the ability to supply parts in the quantity needed by the buyer. Those needs must be clearly stated by the buyer and thoroughly understood by both parties. Delays caused by emergencies that are not controllable by the vendor should be prepared for by the buyer. If the part is unique to the vendor, so that an alternate source cannot be found in a reasonable time, the buyer should maintain a buffer stock for that contingency.

3. *Quickness.* The vendor must demonstrate the ability to react quickly to the needs of the buyer. If the demand for parts increases quickly, the vendor should have a production system with a relatively short flow time that can react to that condition. Ultimately, all customers and vendors should be united in a common goal, constantly striving for a shorter flow time for all activities.

4. *Quote.* The price of the part must be within the guidelines set by the sales department. Many factors must be considered, however, as the final price to be paid for a part by the production system is also based on the other Q's in the list.

It takes about a year to qualify a vendor for 4Q status. That is why so few companies are willing to invest their resources to do it. Once that trust is established, however, the benefits can be profound: higher product quality, lower cost, and superior flow. The following steps can serve as a brief guide to qualifying a vendor for 4Q status:

1. *Plant tour.* A team representing manufacturing, purchasing, QA, materials, and engineering should interview the appropriate personnel at the vendor's plant. All the factors that pertain to vendor qualification should be discussed.

2. *Evaluation.* After the plant tour, the team should discuss and evaluate the vendor. The attitude of the vendor's key people is crucial to success. The qualification procedure is expensive, and it should not be initiated unless success is almost certain. The vendor should be convinced that it is a worthwhile achievement to become qualified. If the vendor fails this evaluation, the team should discuss alternatives.

3. *Follow-up.* The buyer assigned to the vendor should be the single contact source during further negotiations. The buyer should negotiate the agreement between the company and the vendor. The buyer should attend all meetings and discuss the qualification conditions with the vendor.

4. *Conditions.* The conditions of the vendor qualification should be discussed and agreed upon by the team and the vendor. Some of these conditions that apply to the vendor are
    a. Lead time
    b. Operator training
    c. Complete process instructions
    d. Buffer stock
    e. Flow time improvements
    f. Inspection/testing criteria
    g. Test equipment and calibration
    h. Surprise inspections
    i. Employee quality
    j. Quantity prices

5. *Trial period.* The trial period should be open-ended and should terminate when the buyer, in agreement with the team, is convinced that the vendor is qualified. During the trial period, surprise inspections should be conducted by the team at the vendor's plant even if it is a great distance away. Other tests should include 100 percent lot inspections and quick-delivery tests.

6. *Incoming inspection.* The normal incoming inspection procedures should remain in effect during the trial period. Occasionally, the lot should be 100 percent inspected. If a lot from the vendor fails a normal sample inspection, it is a serious breach of the conditions. The vendor should be evaluated again by the team and reconsidered for qualification.

7. *Qualification.* Final qualification should be granted by the buyer and the team. The buyer can then process the paperwork that will allow the vendor and customer to form a closer and more beneficial relationship. Once qualified, the vendor can deliver parts directly to the storeroom, on an open purchase order. In the ultimate situation, parts can be delivered directly to the workbench where they will be assembled.

The above series of steps represents a very brief review of the qualification process. You, as a member of the team, should develop your own methods that will be unique to every vendor under consideration. While you are developing your methods, try to anticipate all the problems that might arise after the vendor is qualified. Then design the evaluation test to solve those problems before they occur.

When you are trying to improve the flow in your plant and are being distracted by poor part quality, it is like trying to carve an eagle on the back porch and being bothered by horseflies. When the horseflies become bothersome, you can shoo them away, but soon they will be buzzing in your face again. You can shoo them away forever, or you can screen in the porch and solve the problem. Qualifying vendors is like putting up screens, so you can concentrate on carving your eagle.

Vendor problems are as old as mass production. When the first moving assembly line was developed by Henry Ford in 1913, he immediately became aware of his dependence on vendors to keep the line moving. Vendor qualification was unheard of, so he bought the vendors. In the end, the Ford Motor Company owned every source of supply for items that had, at some time, shut down the line. In this book, those items are called *critical parts*.

Quality Assurance Inspections

Improvement of this category by means of vendor qualification could result in a major change in the process flow and the layout of your plant. QA inspection stations and other test stations could be eliminated entirely, and unit quality could rise simultaneously. Of primary importance, vendor qualification and operator qualification are prerequisites for a QA inspection reduction program.

These prerequisites exemplify the overall JIT philosophy: All participants of a JIT system should assume responsibility for their actions. This responsibility should extend beyond the point of high-quality processing; it should also include an effort to constantly improve the product. Every inspection and test operation should be investigated to determine how it can be eliminated. Inspection elimination should be initiated on three fronts:

1. *Operator-caused.* If an inspection is required to reveal an operator's mistake, then the operation should be analyzed to eliminate the cause of the error. This may involve more operator training or, better yet, redesigning the unit for error-free assembly. Whatever the reason for inspection, a diligent effort should be made to eliminate the cause or to train the operator to assume more responsibility for the operation.

   A program could be started to certify operators to inspect their own units, similar to the vendor qualification program described before. Each operator should be certified for a variety of operations. The recognition and wages should correspond to the number of operations that the operator is certified to perform.

2. *Part-caused.* Faulty parts are usually not discovered until a unit fails in an electrical or mechanical test. This can be expensive; it is common practice to scrap a printed-circuit

board because it contains a reject component. It is scrapped because the labor cost to find and replace the component is more than the cost of the board.

It is also common to have a component inspected by the vendor and then fail in the assembled unit, because a different testing method was used by the vendor. Any failure condition should be analyzed and resolved, even if the short-term cost seems high. Neglecting to do so can result in a higher product cost, a higher field failure rate, and consequently a reduction in the customer base.

The best prevention plan for part failure in your plant involves creating these test conditions at the vendor's plant:

a. Expose the parts to the same environmental or other test conditions that exist at your plant.

b. Supply the vendor with diagnostic test equipment that exactly duplicates the operating and test conditions used on the finished unit at your plant. The vendor should test the part in exactly the same manner as you do.

c. Agree that the solution to the part malfunction is the vendor's responsibility. That is the JIT philosophy.

If the faulty part was purchased by the vendor and loaded onto an assembly such as a printed-circuit board, then the vendor should trace the problem back to the vendor's vendor. The time involved in organizing such a program is considerable, but the benefits can be immense, and it becomes more routine with repetition.

3. *Process-caused.* This cause of part failure, heat-treating a part, for instance, is the most difficult to eliminate. In the high-tech industry especially, the state-of-the-art processes are seldom perfect. It often takes many years to perfect a sophisticated process, even after the product has been introduced. You must be certain to attack the cause of the problem, not the symptom. Is it really the vendor who is at fault, or is your equipment inadequate for the process? Are the operators lax in their attention to detail, or is the process too demanding for a production operation?

To eliminate any QA inspection operation, the QA group should assume a new, active role in failure analysis. Normally, QA personnel do not investigate the causes of failure. They just inspect the units to ensure that the quality level is satisfactory, then gather and publish failure data at the unit level.

With this new role, QA personnel gather failure data at the unit level, the part level, and the component level. That is, they trace the failure upstream to discover the exact cause, and their failure report reflects that data. The failure report should also include a recommendation by the QA supervisor for eliminating all failures, and consequently eliminating all QA inspection operations.

The QA supervisor is also responsible for operator certification so the operators will be qualified to perform their own inspections. The QA group itself will never be eliminated, however, as there will always be new operators, new products, and new problems. The challenge is to maintain a total high-quality level without QA inspections and with a minimum of staff. It is an ongoing effort, and an entirely different approach to the QA function.

A representative of the engineering department should also assume new responsibilities. If an engineering design function of the product is being tested and units are being rejected, the engineering department should take an active role in the search for the solution, working with QA to determine the exact cause of the rejection. This is a new function for engineering, to become so closely involved with production. As an example, engineering may solve a part quality problem by designing a diagnostic electrical test fixture for the vendor. The production department then supplies this fixture to the 4Q vendor to screen the parts before they are shipped, so they do not have to be inspected when they arrive.

From that time on, when engineering introduces any prototype to production, it should also supply the test equipment that will be needed by the 4Q vendors to test the components. The test equipment should be included in the prototype package when it is transferred to production. That concept breaks new ground for engineering, but is right in line with the JIT philosophy.

All the above effort is directed toward improving production by three different means:

1. Flow time is reduced because rework operations and QA inspection operations are eliminated.

2. Product cost is reduced because no dedicated inspectors are required for the assembly process.

3. Quality is improved because more operations are producing near-perfect units at the source.

Stock and WIP Inventories

This is the category most often associated with JIT. It is related to the expense of carrying excess parts and units. This expense accumulates in the form of interest on the investment in material, floor charges to store the goods, and labor cost to move or update the units. Inventories of all kinds should be reduced to the lowest level that does not threaten production. However, many JIT programs consist only of eliminating obsolete parts from the storeroom, which is a form of housekeeping, not really JIT.

Reducing stock and WIP is so popular among JIT advocates that it is sometimes the only task they consider. Actually, it is a fine-tuning step that should be accomplished toward the end of the program, not at the beginning, and its cost-saving potential is usually minor. Of course, if the system is in complete disarray, this category could be significant, but that is the exception.

Care must be exercised. As an example, one company in the machined parts supply business wanted to initiate a JIT program in the manufacturing and shipping departments. Its line of products was on a no-lead demand basis and had to be shipped the same day as the order was received. The company maintained a large inventory of finished goods and, as a result, had a very loyal customer base. An outside JIT consultant advised the company to cut its finished-goods inventory and then streamline the flow. Read that again: The plan was implemented in the wrong sequence, and it could have put the company out of business!

Qualified vendors are the key to reducing the critical-parts storeroom inventory. Reliable vendors are the key to reducing the noncritical-parts storeroom inventory. Careful planning is the key to reducing the WIP inventory. Streamlining the flow is the key to reducing the finished-goods inventory. Each of these keys must be turned before opening the door to inventory reduction. Of course, minor adjustments in inventories can be made in any system without a formal program, but major changes should be approached with caution. Here is a brief description of these keys to inventory reduction:

1. *Storeroom inventory, critical parts.* This inventory can be reduced only if 4Q vendors supply the critical parts. The qualification process should be tailored to the capability of the vendor. Each part has unique requirements, and each vendor has a different concept of what it means to be qualified.

2. *Storeroom inventory, noncritical parts.* This inventory can be reduced if the vendors are reliable. This evaluation can best be made by the buyer assigned to each vendor. All contingencies that the vendor cannot control must be prepared for. The final inventory quantities should be determined by a cooperative effort among the project leader, storeroom supervisor, production manager, buyer, and vendor.

3. *WIP inventory.* This inventory can be reduced by carefully planning the production process. This book cannot cover all the methods for reducing WIP. A diagram of your process, similar to Fig. 2.3, with queue dots, should serve as a guide. For every group of idle parts, there is a method to eliminate them. The challenge is to devise a method that is cost-effective—some improvements may not be. For example, even if better tooling for quick setups could reduce the WIP, the product may be obsolete before the tooling is paid off.

4. *Finished-goods inventory.* This inventory can be reduced by streamlining the flow. You can streamline the flow by applying all the techniques described in this book. A flowchart similar to Fig. 2.3 should serve as a guide for this project also. A large finished-goods inventory is required only when the product cannot be manufactured quickly. Think about what changes would be required so that your manufacturing system could produce any product within a very short time. Then use that new system as a goal, and decide what improvements will be needed to reach it.

Bottlenecks

The term *bottleneck* is used in this book to describe a restriction in the flow of material through the production process. The alleviation of bottlenecks should be of prime concern when you design a plant layout or flow improvement project. Two of the most common causes of bottlenecks are operator capacity and equipment capacity.

1. *Operator capacity.* Operator capacity is the output of parts from each operation. It can be expressed, for instance, in parts per minute or minutes per part. To balance a series of sequential operations, the output of each operation should be reasonably equal. If they are not, some operators will be idle while others are working feverishly. An automobile assembly operation is a good example of a balanced line with equal operator capacity.

A bottleneck occurs when one operator takes significantly more time than the others to complete an operation. The output of each operation and the output of the line then slow down to the output of the bottleneck. To the casual observer, everyone on the line appears to be equally busy, but in truth, most operators are not. The solution to this dilemma involves operator pace rating and operation work content. This subject will be described in the next chapter.

2. *Equipment capacity.* The equipment capacity bottleneck is similar to the operator capacity bottleneck except that the operation time is controlled by the equipment. Most operations in this category are really controlled by a combination of equipment and operator. The burn-in operation has this dual control, and uses equipment that cannot easily be duplicated.

Because this equipment is so valuable, it must be utilized to the fullest extent. The solution involves assigning more operators to the equipment so it will run constantly. Then it will truly be an equipment-controlled cycle. This subject will be described in the next chapter.

If you would like to read about practical bottleneck solutions, a novel entitled *The Goal*[7] explores the subject in detail. It is a fictional account of a plant manager who saves his division from failure by solving many of its ingrained production problems.

Other Improvements

Value analysis—the evaluation of product costs in terms of contribution to product value—can also be used to improve the product flow. Vendor qualification is a product cost, and better flow time is a product value.

Therefore, the designers of new products should evaluate the sources of their critical parts and ensure that all of them are either qualified or potential 4Q vendors. The designers may even have to visit some of these vendors before approving them.

When a product is transferred to production, the package should include a list of critical parts along with the selected vendors. Again, that is the JIT philosophy.

Other improvements that can be included in a layout form an endless list that may or may not pertain to your plant. The items shown in Fig. 3.1, a list of improvements, will apply to most plants. Examine this list for items that relate to your plant, and then add more as they come to mind.

4

Labor Analysis

In this chapter, the direct labor used by the Flexible Division to assemble its Memory Bank Model 107 will be analyzed. A mathematical model of the labor content will be constructed and used to justify the improvements that the IE wanted to make.

CEDAR Analysis

The spreadsheet used in this book to analyze the production labor is part of a total product cost analysis that has been developed by the author. The complete spreadsheet is called the *CEDAR* (Cost Evaluation Data And Report) *Analysis.* It is beyond the scope of this book to describe the entire CEDAR Analysis in detail, but the labor section makes an excellent starting point for a flow improvement project. Every section of the CEDAR Analysis is a mathematical model of the corresponding part of the production system. The labor section, by itself, has a variety of uses:

1. To balance the operation times on an assembly line.
2. To determine the amount of labor or time that can be saved by making improvements in the system.
3. To analyze and improve the flow time.
4. To locate the bottleneck.
5. To calculate the capacity of certain equipment used in the operation.
6. To provide the data for pie charts or other visual aids that may be helpful to support an improvement of the system.

If your manufacturing system is very complex with many different machines serving a variety of work orders, you should consider a simulation program. A busy job shop is a good example, where a punch press may become overloaded with work orders if the arrivals are not closely timed. For the Flexible Division, however, the IE chose a simple spreadsheet because it was fast and accurate, and provided a realistic analysis of the process.

In this book, the term *old* refers to the Flexible Division manufacturing system as it existed about a year before it was moved to the new plant. At that time, the market had grown very rapidly, and the future looked bright for its unique product, the RAM/ROM Memory Bank. As a result of this rapid growth, the assembly area was not well arranged, no vendors were qualified, and the storeroom was so disorganized that only the clerk knew where the parts were.

The term *new* refers to the manufacturing system that the IE planned for the new plant layout. This new system incorporates all the improvements shown in Fig. 3.1, the list of improvements, and many more. Each old condition, before the improvement was made, is listed in this chapter under the heading Old Flexible Division and incorporated into the old CEDAR Labor Analysis. Each corresponding improvement is listed under the heading New Flexible Division and incorporated into the new CEDAR Labor Analysis.

The old and new systems are then compared, and the savings are analyzed under the heading Time Comparison. This entire chapter is an example of how the CEDAR Analysis can be used to improve the manufacturing system and justify those improvements.

In both of the spreadsheets, most figures have been programmed for a one-place decimal format. The actual value still resides in the cell and is used in all calculations, but is not shown on the printout.

Old Flexible Division

When the Flexible Division started producing the RAM/ROM Memory Bank, it was in the enviable position of having more customers than the production area could handle. The assembly area was set up by simply moving equipment into the first available space, and support functions were located nearby. The units were often rejected for mechanical assembly reasons, and the stack of reject circuit cards was becoming unmanageable. An industrial engineer (the IE) was hired to reorganize the manufacturing system and to introduce JIT into production. A new plant layout that would facilitate the reorganization would be needed later, as the company was planning to move in about a year.

As a first step, the IE made Fig. 2.2, the old process flowchart, and Fig. 2.3, the new process flowchart, to document the flow and to become acquainted with the assembly process. Then the IE prepared this list of unfavorable conditions which should be improved to streamline the flow. (It corresponds to Fig. 3.1, the list of improvements.)

1. *Vendors.* No vendors were qualified. Parts were often delivered late. Circuit cards often failed at electrical test and burn-in. Some lots composed of critical parts were rejected at incoming inspection, but rather than return the lot, the good parts were screened out and used immediately.

2. *Receiving.* The shipping and receiving areas were combined, and packages were occasionally processed incorrectly because of that condition. New parts that were supposed to be sent to incoming inspection were sometimes treated as reject parts and were sent back to the vendor.

3. *Incoming inspection.* The flow through the area was erratic. Boxes were lost in the shuffle, and panic searches were common. The area could not be closed off, so production workers often helped themselves when they needed parts in a hurry.

4. *Parts issue.* In the storeroom, all the parts for six units were loaded onto two carts, wheeled out, and placed in an empty area on the production floor. During rush periods, the aisles were lined with carts.

5. *Storeroom.* Parts were stocked in a random order. If the clerk was absent, a team of assemblers had to find the parts to make a stock issue for a work order.

6. *Assembly.* The assembly area was arranged as a job shop. Each bench performed only one assembly operation. The product was made in five different models, and each model had a different sequence of operations. Confusion was common, and often units were assembled incorrectly and had to be reworked.

7. *Inspection.* A QA inspector was on the floor full-time because the operators were not motivated to take responsibility for their work. It was common for the QA inspector to find mechanical errors in the assembly.

8. *Electrical test.* Not one unit passed the first electrical test. Usually there were multiple causes of the rejection, but no one was sure why many of the units failed. No failure analysis program was in effect. Reject circuit cards that could not be easily repaired were stacked on shelves with little hope that they would ever be salvaged.

9. *Electrical test equipment.* The test bench was covered with a variety of testing gear, used for analyzing the circuit cards. Troubleshooting was required at the component level, so senior technicians were needed at all times. These technicians were frustrated by the repetitious work. The station operated for several hours overtime and was still short of capacity.

10. *Burn-in.* Over 30 percent of the units failed at the burn-in operation, even after passing the electrical test. The burn-in was made at room temperature, for 48 hours, using a functional operating cycle. Some units failed at the end of the test, so the time for all units was extended to 72 hours. More burn-in racks were on order. All units were tested 24 hours a day, but were turned off on weekends.

11. *Finished-goods inventory.* The sales department insisted on having a full line of products in inventory at all times, because the demand for any one product could not be accurately

predicted. Many products were sold because the salesperson could guarantee immediate delivery. As a result, the finished-goods inventory was always overloaded with some products and short of others. The investment interest on this expensive inventory was a constant source of irritation between the sales and accounting departments.

12. *Thick-film process.* A small number of microcircuits failed in the field because the thick-film resistors drifted out of tolerance with time. Customers complained when they were denied access to their data banks because the owner code entry was not recognized.

13. *Status meetings.* These meetings were held every morning for about an hour. Few problems were ever solved, but the production schedule was constantly adjusted when work order priorities were changed. Lately, the discussion centered on buying an expensive bar-code system that would keep better track of the work order status on the production floor. In that way, maybe the meetings could be held less often.

The above list of unfavorable conditions may imply that the Flexible Division operation was out of control. It was not. On the contrary, it was typical of a company that sells a product on the leading edge of its industry. There are many companies that operate today under similar conditions and are still very profitable. Imagine, however, how they could dominate their field if they were properly organized.

Old CEDAR Labor Analysis

Next, the IE used the CEDAR Labor Analysis to document the existing conditions of the production system. This spreadsheet is shown in Fig. 4.1 and is contained in the file named LABOR.WK1 on the diskette. It is a Lotus 1-2-3 file and contains both Fig. 4.1 and Fig. 4.2, the new CEDAR Labor Analysis. If you want to construct your own CEDAR Labor Analysis spreadsheet from scratch, all the formulas are described in Fig. 4.3.

Notice that Fig. 4.1, the old CEDAR Labor Analysis, is based on Fig. 2.2, the old process flowchart. The operation numbers, the reject rates, and other data on the flowchart coincide with their counterparts on the spreadsheet.

On the spreadsheet shown in Fig. 4.1, the upper section contains some constant values that the IE entered. These values are used in the calculations in the lower, tabular section of the sheet. Instead of using these values, you can substitute values that pertain to your operation, and even add others unique to your system. Two open lines have been included for these entries in the spreadsheet. Then you can modify the tabular section to use these new values.

There are several advantages to having the constant values of the system displayed at the top of the spreadsheet:

DISKETTE FILE: LABOR.WK1
PIE CHART NAME: OLD-PIE

15% = PERSONAL, FATIGUE AND DELAY ALLOWANCE.
72 = HOURS, BURN-IN CYCLE.
6.7 = MINUTES/UNIT TO PREPARE PARTS IN STOREROOM: 20 MINUTES PER CART   2 CARTS NEEDED FOR   6 UNITS

147 = UNITS PER WEEK, DEDICATED OUTPUT, MODEL 107 MEMORY BANK

QUEUE HRS	OPERATIONS	REJECT RATE	REWORK --> MIN/UNIT	NORMAL MIN/UNIT	HRS/UNIT	STAND HRS/UNIT	NUMBER PEOPLE, STATIONS	UNITS/HR	HRS/DAY	UNITS/DAY	BOTTLE NECK
2.0	PACK PARTS, STOREROOM			6.7	0.1	0.1	0.5	3.9	8	31.3	*
4.0	#1 ASSEMBLE CHASSIS			12.5	0.2	0.2	1.0	4.2	8	33.4	*
4.0	#2 MOUNT BACK PLANE			12.0	0.2	0.2	1.0	4.3	8	34.8	*
3.0	#3 BUILD CARD CABLE	5%	10	11.0	0.2	0.2	1.0	4.5	8	36.3	*
2.0	#4 BUILD MAIN CABLE	5%	10	12.0	0.2	0.2	1.0	4.2	8	33.4	*
2.0	QA CABLE TEST			8.0	0.1	0.2	0.6	3.9	8	31.3	*
4.0	#5 CONNECT CABLES	4%	8	10.5	0.2	0.2	1.0	4.8	8	38.6	*
2.0	QA WIRING TEST			6.0	0.1	0.1	0.6	5.2	8	41.7	*
4.0	#6 TEST CARDS	10%	5	10.0	0.2	0.2	1.0	5.0	8	39.8	*
3.0	#7 FINAL ASSEMBLY			8.0	0.1	0.2	1.0	6.5	8	52.2	*
5.0	#8 ELECTRICAL TEST	60%	16	17.0	0.4	0.5	1.5	2.9	10	29.4	<-- HERE
5.0	#9 BURN-IN TEST	30%	20	20.0	0.4	0.5	1.5	3.0	10	30.1	*
3.5	QA INSPECTION	10%	10	10.0	0.2	0.2	0.8	3.8	8	30.4	*
1.0	FINISHED GOODS STOCK			5.0	0.1	0.1	0.4	4.2	8	33.4	*
0.0	PACK AND SHIP			6.0	0.1	0.1	0.5	4.3	8	34.8	*

EQUIPMENT OPERATIONS

| 0.0 | #9 BURN-IN RACK | | | | | | 120 | 1.4 | 24 | 34.8 | * |

45 = TOTAL QUEUE HOURS   TOTAL STANDARD HOURS =   83   3.3

PIE CHART DATA

FLOW TIME	PIE CHART DEGREES
3 WORK HOURS	5
83 BURN-IN HOURS	132
45 QUEUE HOURS	71
96 OFF HOURS	152
227 HRS FLOW TIME	360 DEGREES
9.5 DAYS FLOW TIME	2 IN. RADIUS

Figure 4.1 Old CEDAR labor analysis.

O
66
67
68
69
70
71
72
73
74
75
76
77
78
79
80
81
82
83
84
85
86
87
88
89
90
91
92
93
94
95
96
97
98
99
100
101
102
103
104
105
106
107
108
109
110
111
112
113
O

DISKETTE FILE: LABOR.WK1
PIE CHART NAME: NEW-PIE

15% = PERSONAL, FATIGUE AND DELAY ALLOWANCE.
60 = HOURS, BURN-IN CYCLE.
5.5 = MINUTES/UNIT TO PREPARE PARTS IN STOREROOM: 11 MINUTES PER CART   6 CARTS NEEDED FOR 12 UNITS

155 = UNITS PER WEEK, DEDICATED OUTPUT, MODEL 107 MEMORY BANK

QUEUE HRS	OPERATIONS	<-- REWORK --> REJECT RATE	MIN/UNIT	MIN/UNIT	HRS/UNIT	STAND HRS/UNIT	NUMBER PEOPLE, STATIONS	UNITS/HR	HRS/DAY	UNITS/DAY	BOTTLE NECK
1.0	PACK PARTS, STOREROOM			5.5	0.1	0.1	0.5	4.7	8	37.9	*
0.5	#1 ASSEMBLE CHASSIS			12.5	0.2	0.2	1.0	4.2	8	33.4	*
0.5	#2 MOUNT BACK PLANE			12.0	0.2	0.2	1.0	4.3	8	34.8	*
0.5	#3 BUILD CARD CABLE	1%	10	12.0	0.2	0.2	1.0	4.3	8	34.5	*
0.5	#4 BUILD MAIN CABLE	1%	10	13.0	0.2	0.3	1.0	4.0	8	31.9	*
0.5	#5 CONNECT CABLES			11.5	0.2	0.2	1.0	4.5	8	36.3	*
0.5	#6 INSERT CARDS										
0.5	#6 FINAL ASSEMBLY			13.5	0.2	0.3	1.0	3.9	8	30.9	<-- HERE
0.5	#7 BURN-IN TEST	10%	15	20.0	0.4	0.4	1.5	3.6	9	32.8	*
1.0	FINISHED GOODS STOCK			5.0	0.1	0.1	0.4	4.2	8	33.4	*
0.0	PACK AND SHIP			6.0	0.1	0.1	0.5	4.3	8	34.8	*
	EQUIPMENT OPERATIONS										
0.0	#9 BURN-IN RACK						120	1.9	24	45.7	*
6.0 = TOTAL QUEUE HOURS	TOTAL STANDARD HOURS =				63	2.2					

TIME COMPARISON	OLD	NEW	DIFF.	%
LABOR HOURS	3.3	2.2	-1.2	-35%
FLOW TIME HOURS	227	87	-140	-61%
FLOW TIME DAYS	9.5	3.6	-5.8	-61%
OVERTIME HOURS	4.0	1.0	-3.0	-75%
OUTPUT, UNITS PER WEEK	147	155	7	5%

PIE CHART DATA

FLOW TIME	PIE CHART DEGREES
	9
2 WORK HOURS	9
63 BURN-IN HOURS	259
6 QUEUE HOURS	25
16 OFF HOURS	67
87 HRS FLOW TIME	360 DEGREES
3.6 DAYS FLOW TIME	1.2 IN. RADIUS
=====	

Figure 4.2 New CEDAR labor analysis.

1. It is convenient to identify what constants the spreadsheet is based on and what their value is.

2. A very quick "what if" comparison can be made by changing the constants and noticing the results in the summary section at the bottom of the spreadsheet.

3. The derivation of the constant can be shown. For example, the "minutes/unit to prepare parts in storeroom" is calculated and displayed on the same line with the constant.

The "units per week," however, is a calculated value, based on the bottleneck restriction of capacity. That is, the capacity of the system is the same as the capacity of the bottleneck. This value is calculated in the tabular section of the spreadsheet, and then it is displayed as "units per week" in the constants section. The description of these constant values are as follows:

PERSONAL, FATIGUE, AND DELAY ALLOWANCE. The IE entered this typical value of 15 percent, but you may wish to modify it to suit your industry standard. The *Handbook of Industrial Engineering*[1] (p. 1628) shows other examples: 15 percent, bench assembly; 16 percent, floor assembly; 17 percent, turret lathe; 21 percent, drop forge.

HOURS, BURN-IN CYCLE. The IE entered this burn-in time for the memory bank. It serves as an example of equipment that controls a cycle in the assembly process.

MINUTES TO PREPARE PARTS. The IE entered values for the time to load the cart, the number of carts needed, and the number of units per cart and calculated this value from them.

UNITS PER WEEK. This value is calculated from the capacity of the bottleneck, shown in the tabular section.

The descriptions of the column titles are as follows:

QUEUE HRS. Time, in hours, that the unit waits after each operation has been completed.

OPERATIONS. A list of the operations copied from Fig. 2.2, the old process flowchart. You can change these descriptions to suit your flowchart.

REWORK REJECT RATE. The average percentage of units that are rejected and reworked at any operation.

REWORK MIN/UNIT. The average time in minutes that it takes to rework a rejected part.

NORMAL MIN/UNIT. Raw operation time in minutes, including setup, paperwork, cleanup, etc., converted to normal time by pace rating.

HRS/UNIT. Converts normal minutes per unit to normal hours per unit.

STAND HRS/UNIT. Adjusts normal hours per unit for the personal, fatigue, and delay allowances shown in the constant values section, and results in standard hours for the operation.

NUMBER PEOPLE, STATIONS.   Number of people or stations dedicated to the operation.

UNITS/HR.   Total output of the station in units per hour.

HRS/DAY.   Number of hours per day that the station is active.

UNITS/DAY.   Total output of the station per day.

BOTTLENECK.   Indicates what operation is creating the current bottleneck. The operation is indicated by the <--- HERE arrow. It automatically moves to the proper location as you revise your data.

The first step in filling out the labor analysis spreadsheet is to list the operations in sequential order under the heading OPERATIONS. Then start gathering data that will complete the analysis. Before you go into the assembly area to start gathering data, check out your plan with the supervisor. You may have to interrupt the operators for some time during the process, and the supervisor will want to know why you are there. The supervisor may even want to inform all the operators of your presence and function before you start, especially if it is a union shop.

The QUEUE HRS can be determined by observation and timing. The rework REJECT RATE and MIN/UNIT can be estimated by the operators. Most data of this type does not have to be precise, and it can be made more accurate later if necessary.

If no standard data is available for the operation times, then the MIN/UNIT will have to be determined by time study. This can be brief, but you should still arrive at a raw time that accurately represents the time it takes to perform the operation. Be sure to prorate the time required for the setup, cleanup, paperwork, etc., involved.

Normalize the raw time by pace rating the operator. If you have no experience with pace rating, find someone in your plant who has, or study the pace of as many operators as possible to arrive at the activity level of a normal (100 percent) pace. Then become familiar with what the various percentages of normal pace look like in action. In the end, you should be able to observe an operator and estimate with some degree of accuracy what percentage of normal pace is being performed.

This ability to pace rate operators is especially important if your system is an assembly line. Ideally, everyone should be working at a normal pace, but not everyone is normal or is feeling normal every day. There will be a bottleneck somewhere on the line, and all operators will unconsciously pace themselves to it, unless the line is mechanically timed like an automobile conveyor.

The NUMBER PEOPLE, STATIONS and HOURS/DAY can be derived from the operators. The UNITS/DAY is a calculated value. The BOTTLENECK is automatically located by the spreadsheet every time you make a change that affects the output of the system.

The purposes of this analysis for a plant layout project are many:

1. *Flow time.*   The spreadsheet lists all the elements that comprise the total flow time. You can now focus on each element

and determine how the queue hours can be reduced or eliminated. As you improve the queue time, the shorter flow time is shown at the bottom of the analysis, expressed in working days.

2. *Rework.* The rework operations become very obvious when they are listed in their own column. However, reducing or eliminating them requires a dedicated effort. This involves operator training, qualifying 4Q vendors, and designing more sophisticated test equipment that is easier to operate.

3. *Operation times.* As you time-study each operation, many improvements may become obvious, especially if the operations have never been timed before. This is your chance to make them more efficient, although it could be time-consuming. Be sure to update the operation instructions also. The number of stations may be affected, and that will definitely have an impact on the layout.

4. *Bottleneck.* This item is important, since the bottleneck will set the pace for the entire production line. Simply by shifting some labor hours away from the slow-paced operations into the bottleneck operations, you can increase the output of the system at no cost at all. Fine-tune the operation until all the UNITS/DAY are within the same general range.

5. *Machine loading.* Some machines are very expensive and must be utilized to the fullest. One way to do this is to assign extra operators to the equipment. One person can start an hour early in the morning, then trade off for breaks and lunch, and the other person can work an hour later at night. They both work only eight hours, but the machine is being used for ten hours, without interruption. Even if the idle person has no other duties, this procedure could raise the output of the entire production system significantly under the right circumstances. An important bottleneck could be alleviated, and the improvement would appear on your spreadsheet.

New Flexible Division

The IE studied the list of unfavorable conditions mentioned earlier in this chapter and then made some assumptions, based on the fact that the improvements could be made before the new plant layout was designed. The IE made a new list, showing what could be accomplished if each condition were resolved successfully. In this new list, each condition has been developed and fully implemented.

1. *Vendors.* All vendors for critical parts were qualified as 4Q vendors. Quality at the unit level improved dramatically, and the need for QA and electrical testing was drastically reduced.

2. *Receiving.* The shipping and receiving areas were separated to avoid mixing of the two categories.

3. *Incoming inspection.* The flow through the area was controlled by a conveyor, and the area was walled off for security.

4. *Parts issue.* In the storeroom, all the parts for each operation, not each product as before, were loaded into bins on carts, and wheeled out into an allocated space by a bench.

5. *Storeroom.* The parts were stored in subassembly order wherever possible. The attendants had no trouble memorizing the new part locations.

6. *Assembly.* The assembly area was arranged as an assembly line for flexible manufacturing of all products. Any one of five products could be started on demand within 15 minutes' notice, and completed in about four days.

7. *Inspection.* All QA inspections were eliminated. The QA supervisor became involved in every aspect of quality improvement and eventually qualified all operators for self-inspection.

8. *Electrical test.* The product was formally tested only once, and that was after burn-in. All other formal tests were eliminated after they were monitored and found to produce an insignificant number of failures. Some operators tested their own units very briefly.

9. *Electrical test equipment.* The testing equipment used by the operators for brief testing was redesigned for diagnostic operation. The few units that failed were repaired on the spot by the operator. No unit was ever scrapped.

10. *Burn-in.* The burn-in cycle was reduced from 72 to 60 hours, after an insignificant number of units failed past 60 hours. The reject rate was reduced from 30 to 10 percent. The order for more burn-in racks was canceled.

11. *Finished-goods inventory.* The inventory was almost eliminated because the flow time had decreased from two weeks to less than one week.

12. *Thick-film process.* The abrasive trimmer for resistors and capacitors was replaced by a laser trimmer. The laser trimmer was able to achieve a closer tolerance on the resistor and capacitor values, and the component drift problem all but disappeared.

13. *Status meetings.* The status meetings were eliminated. Any problems were handled on an individual basis instead. The bar-code equipment that had been considered for tracking production work orders was not needed.

New CEDAR Labor Analysis

Next, the IE copied the file of Fig. 4.1 and revised it to produce Fig. 4.2, the new CEDAR Labor Analysis. The IE assumed that all the changes proposed in the preceding list had been accomplished. This Fig. 4.2 is based on Fig. 2.3, the new process flowchart. Many significant advantages resulted from these changes that the IE made in the manufacturing system:

1. *Burn-in cycle.* The burn-in cycle was reduced because the 4Q vendors were testing the components more realistically and thus were supplying a higher-quality part tailored to the use of the Flexible Division.

2. *Prepare parts in storeroom.* This unit time decreased because the parts were stored in kit picking order. Parts that were common to several kits were stored separately.

3. *Units per week.* This increased as a result of the improvements incorporated into the system.

4. *Rework.* All the inspection operations were eliminated as a result of the improved part quality and the QA supervisor's inspection reduction program. The reject percent of the remaining test operations also improved.

5. *Minutes per unit.* Many of these unit times were changed as a result of combining some operations and eliminating others.

6. *Hours per day.* The overtime was reduced considerably.

7. *Bottleneck.* The bottleneck shifted from the electrical test to the final assembly operation, but more important, all the operations were fine-tuned to produce an even pace for all operators.

Time Comparison

Although the example used in the CEDAR Labor Analysis is a high-tech product, the principles are universal and can be applied to virtually any manufacturing plant. They can even be used for analyzing a warehouse or order-picking facility, or any operation that is concerned with material flow.

You can adjust the spreadsheets to illustrate how any product is affected by the chemical, physical, mechanical, electrical, or material-handling functions involved. Your vendors can be qualified to supply material that conforms to your desired properties, not necessarily to their usual final test procedure.

The contents of the spreadsheet cells are described in Fig. 4.3. Not every cell is'described, but the general formula for a category of cells is given. If the data was supplied by the IE, it is referred to as "IE entered data."

In the lower left-hand corner of Fig. 4.2 is the TIME COMPARISON section of the spreadsheet. Each category of savings that resulted from the new CEDAR Labor Analysis is presented as a difference and as a percentage:

TIME COMPARISON	OLD	NEW	DIFF.	%
LABOR HOURS	3.3	2.2	–1.2	–35%
FLOW TIME HOURS	227	87	–140	–61%
FLOW TIME DAYS	9.5	3.6	–5.8	–61%
OVERTIME HOURS	4.0	1.0	–3.0	–75%
OUTPUT, UNITS PER WEEK	147	155	7	5%

The formulas shown below are for the entries in the first cell of each range. The formula in the second cell of the range will be relative to the first cell. For instance, in the first item, the cell entry for A6 is @CELL("ROW",A6..A6). The cell entry for A7 is @CELL("ROW",A7..A7), and so on.

Cell descriptions for Figure 4.1:

A6 to A52 = @CELL("ROW",A6..A6)

This formula assigns numbers to the rows of the spreadsheet for demonstration purposes and will not be needed for your analysis.

B6 = Allowances

IE entered data. It totals the percentage of time normally allowed for personal time (5%), miscellaneous delay time (5%), and fatigue slowdown time (5%).

B7 = Burn-in cycle

IE entered data. It is the burn-in cycle time for each unit.

B8 = +G8*J8/M8

This formula calculates the time required by the storeroom personnel to prepare the parts for the assembly operation for one unit.

B11 = @MIN(L18..L36)*5

This formula calculates the output of the assembly area for a 5-day week, based on the bottleneck output. As the bottleneck changes from one operation to another, the output changes.

B18 to B36 = Queue hours

IE entered data. It is the amount of time that each unit waits for the next operation. Assembly time is not included.

B38 = @SUM(B17..B37)

This is the sum of all queue hours.

C18 to C32 = Labor operations list

IE entered description. It is a list of all labor operations required to build the product.

C36 = Equipment operations list

IE entered description. The equipment in this list controls part or all of the cycle time of the operation.

D18 to D32 = % reject rate

IE entered data. It is the percentage of units that are rejected and then reworked at each station.

E18 to E32 = Rework minutes per unit

IE entered data. It is the average normal time used to rework the unit after it has been rejected.

F18 = +B8

F19 to F32 = Minutes per unit normal time

IE entered data. The IE performed a brief time study of each operation. This figure represents the average time to complete each operation, including the prorated time needed for setup, cleanup, paperwork, and other necessary functions. The IE estimated the performance rating of the operator and multiplied the operator time by the performance rating to arrive at the minutes per unit normal time. Example: 10 minutes operation time $\times$ 90% performance rating = 9 minutes of normal time.

Figure 4.3 Cell description.

G8 = Minutes per cart

IE entered data. It is the time required by the storeroom operators to load an average cart with parts.

G18 to G32 = ((D18*E18)+F18)/60

This formula adds the prorated rework time to the minutes per unit normal time and converts the total to hours per unit.

G36 = (1+(D29/2))*B7

This formula calculates the average time required to burn in each unit. The burn-in time is multiplied by 1 plus the average reject rate to arrive at the average time to burn in each unit. The reject rate is divided by 2 because the average reworked unit additional burn-in time is only half a cycle.

H18 to H32 = (1+B$6)*G18

This formula adjusts the hours per unit for the allowances, and the result is the standard hours per unit.

H38 = @SUM(H17..H37)

This is the sum of all standard hours.

I18 to I36 = Number of people or stations

IE entered data. It is the number of people or stations that are dedicated to this product while it is being assembled. Fractions of people are used when they have other duties that are not connected with the production run. The burn-in figure refers to the number of bays in the burn-in rack.

I46 = @ROUND(H38,0)

I47 = @ROUND(G36,0)

I48 = @ROUND(B38,0)

I49 = (I46+I48)*2

This formula calculates the nonwork hours that are lost at night. The hours used during the day shift are multiplied by 2, for the two shifts that are lost at night.

I51 = @SUM(I46..I50)

This is the sum of all flow time in hours.

I52 = +I51/24

This formula converts the flow time in hours to days, and the result is the total flow time of the product, expressed in days.

J8 = Number of carts with parts

IE entered data. It takes this many carts to hold a set of parts that are needed to make the number of units shown in M8.

J18 to J32 = 1/H18*I18

This formula calculates the number of units per hour that can be produced in this operation by the number of operators shown in I18.

J36 = 1/G36*I36

This formula calculates the output of the burn-in rack in units per hour.

Figure 4.3   Cell description (*Continued*).

K18 to K36 = Hours per day

    IE entered data. This is the number of hours per day that each station is operating.

L18 to L36 = +J18*K18

    This formula calculates the number of units per day produced by each station.

L46 to L49 = +I46/I$51*360

    This formula calculates the number of degrees in the pie chart that are assigned to each category of flow time.

L51 = @SUM(L46..L50)

    This is a check sum and should be 360 degrees.

L52 = Radius of the pie chart

    IE entered number. This figure was chosen by the IE for the radius of the pie chart shown at the top of Fig. 4.3.

M8 = Number of units

    IE entered data. Parts for this many units are contained in the set of carts shown in J8.

N18 to N36 = @IF(@MIN(L$18..L$36) = L18,"<--HERE","   *")

    This IF statement tests column L for the minimum value and then displays a <--HERE sign on the lowest number shown on that row. The result is a moving indicator that highlights the operation with the lowest output, which is the bottleneck. The asterisk indicates which cells contain the bottleneck formula.

Cell descriptions for Figure 4.2:

The cell descriptions in Fig. 4.2 are the same as those for Fig. 4.1 except for the lower left-hand section, entitled TIME COMPARISON. These are the cell descriptions for that area:

C106 = Comparison categories

    IE entered description. It is a list of categories that can be compared by using the old and new labor analysis.

D106 = +H38

D108 = +I51

D109 = +I52

D111 = @SUM(K18..K33)-(@COUNT(K18..K33)*8)

D113 = +B11

E106 = +H98

E108 = +I111

E109 = +I112

E111 = @SUM(K78..K93)-(@COUNT(K78..K93)*8)

E113 = +B71

F106 = +E106-D106

G106 = +F106/D106

Figure 4.3 Cell description (*Continued*).

There are significant savings in all categories, but one stands out as the dominant force driving your layout: the flow time. The flow time is the most important attribute of a manufacturing system that you should analyze before you design a new plant layout. Notice that the flow time was reduced by 61 percent of its original value. That reduction is so great that it will affect the way the company operates in the future, and it also affects the plant layout dramatically:

1. The number of work stations has been reduced.

2. The bench arrangement has changed from a job-shop to an assembly line sequence.

3. The units now travel between work stations on a conveyor.

4. The number of carts has been reduced.

5. The storeroom area can be reduced.

The flow time pie charts in Fig. 4.4 illustrate the improvement gained in each category affected. The total areas of the pies are proportional to the flow time. The *off hours* refer to the time that the plant is closed. In this case, since it is a one-shift operation, the off hours are 16 hours per day. The total hours of the new flow time can be added up as though they occurred in sequence:

2 hours of work time

6 hours of queue time

16 hours when the plant is closed

63 hours of burn-in time
———
87 hours total flow time

Notice how much the queue hours sector has been reduced as a proportion of the total flow time in the new system. It is also obvious that the next attempt to improve the flow time should concentrate on reducing the burn-in time.

Figure 4.4 was constructed by using AutoCAD in order to size the pies properly. The data used to construct them is shown in the lower right-hand corner of Fig. 4.2 under PIE CHART DEGREES. The drawing of the two pie charts is included on the diskette as a file named PIE.DWG.

Both of these pie charts are also included in the Lotus spreadsheet file on the diskette. The names of the pie charts, OLD-PIE and NEW-PIE, are shown at the top of each spreadsheet. However, these two pie charts will be the same size if you print them.

This chapter concludes the analysis phase of the plant layout project. The improvements made by the Flexible Division IE will be referred to often, and the implementation of these improvements will be explained in detail. But for the most part, the rest of the book will be concerned with the actual design of the plant layout.

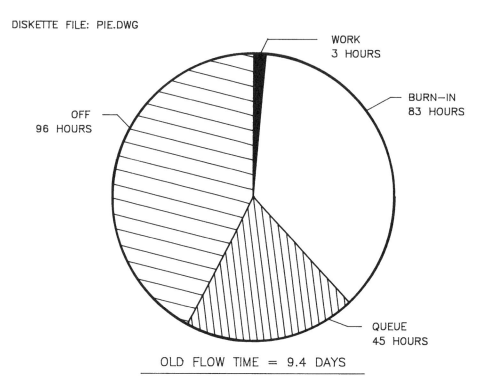

OLD FLOW TIME = 9.4 DAYS

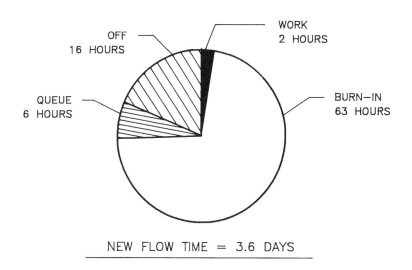

NEW FLOW TIME = 3.6 DAYS

AREA OF PIE CHART IS PROPORTIONAL TO FLOW TIME

Figure 4.4   Flow time pie charts.

5

New Plant Planning

Criteria to aid you in the selection of a building for a new plant will be provided in this chapter. This selection will be based, for the most part, on the equipment list, which determines how much floor space will be required for the new production system. Other considerations will be covered at the end of the chapter.

Floor Space Planning

If you are going to move an existing system with no changes in the assembly process, the area now occupied will be a good estimate of the area that will be required. More space will be needed, however, if the present area is crowded, if some areas will be divided and separated, or if some new areas are going to be added to the new plant layout.

In Fig. 4.2, the new CEDAR Labor Analysis, each operation was listed in assembly sequence. This is an excellent first step toward floor space planning for the production area, but much more organization is still required. The equipment to accomplish those operations must be determined, listed, and measured before the area allocation can begin.

If you are going to assemble a prototype product, it can be frustrating to decide what equipment will be needed. Also, you will have to arrange this equipment for a process flow that has only been performed in the prototype lab. Those processes may not represent the ones that will be used in production, yet the space allocated for the system may be frozen after you make your layout. You may be in the position of planning for a process that will need twice as much room in the near future, while not knowing what the actual requirements will be.

If that is your situation, this is the time for a *just-in-case* (JIC) plan. Not a joke, JIC is the antithesis of JIT. When you use a JIC plan, numerous problems are expected, parts are ordered well in advance and are inspected to death, inventories swell, the floor area is spacious, and plans are in place for rapid crisis resolution. It is often used in manufacturing military products when the production runs are too small to develop the experience necessary for careful JIT planning. JIC can be a legitimate survival plan if you are assigned a product that had minimal production planning and contains some state-of-the-art processes.

The best defense for this situation is to assign some blank areas in your layout for future, custom-designed equipment that will accomplish the same processes, but in a more efficient, reliable manner. The old equipment must remain in place to support current production until the new equipment can be built, debugged, and placed on line. Leave room for more benches or other equipment that may be needed; otherwise, they will crowd your area and make production impossible. A good JIC plan will grant you the freedom to make major changes in your layout and still incorporate them into the overall plan.

However, the more common requirement for a plant layout is to accommodate a new production plan or to expand the capacity of an existing system. The Flexible Division is moving to a new plant because the old plant does not provide the amenities of a modern facility, and a new production system and layout are badly needed. Since the production processes have been in place for some time, the IE is sure of the equipment needs, and only a few new pieces will be required to provide more flexibility.

The IE also designed the assembly line to accommodate future products. An extra bench will be added so if future products require more assembly operations, the layout will not have to be changed.

Be sure to allow room for expansion in your preliminary plans, especially if work centers are going to be walled off. An example would be a production support operation like the Flexible Division thick-film area. The IE added another conveyor furnace to this area so the existing furnace would not have to be adjusted to a new profile every time the conductors were fired. A laser trimmer was added to provide a more reliable product. These two additions dramatically expanded the capacity of the work center, allowing it to supply the parent corporation with all its thick-film needs. If you have a similar operation that has just enough capacity to supply production, then assign more floor space to that work center in the new building.

To start the planning phase of this step, first make a brief tour of the entire existing facility. Notice how many general areas the present plant is divided into, and consider how these areas will fit into the new plant. Notice how much equipment is located in each area. How much of it is standard, such as benches, shelving, file cabinets, etc., and how much is nonstandard, such as large machinery, storage bins, or custom assembly equipment? Bear in mind that every item that will be moved to the new location must be entered on the equipment list and described in these terms:

1. *Numbers.* Every piece of equipment, including large trash cans, will be assigned a number in a series that is unique to the area where the equipment is located before the move.

2. *Dimensions.* Every piece of equipment will be measured for width, depth, and height.

3. *Utilities.* Each piece of equipment that has a utility requirement will have it noted on the list, including a 110-V outlet for each workbench.

This detailed documentation may seem like overkill as you first start the data collection process, but the amount of advance planning that you do now will pay off handsomely during the move. Moving day can be a horrendous, disorganized calamity that your company may long remember and never fully recover from. Or it can be a smooth transition that everyone forgets soon after they have settled in. To a great extent, you make that choice when you decide how much pre-planning to do.

Equipment List

The IE for the Flexible Division designed the spreadsheet shown in Fig. 5.1, the new plant equipment list, to organize and describe every item in the old plant that will be moved and the new equipment that will be purchased. This spreadsheet is on the diskette as a Lotus 1-2-3 file named LIST.WK1. It summarizes the equipment requirements of the plant, and calculates the area needed for each work center in the new plant. The cell descriptions for this spreadsheet are listed in Fig. 5.2.

The term *footprint* is defined as the area of the equipment that is resting on the floor, including low protrusions that are obviously part of that area. An *overhang* is a part of the equipment that extends out at some distance above the floor and interferes with access to the equipment. Although the knee space of a desk is technically an overhang, it is still considered part of the footprint. In the spreadsheet, the width and depth of the equipment include the footprint and the overhang.

Notice the expansion factors listed at the top of the sheet in the constants section. They are different for each work center. The actual floor area (the footprint plus overhang) of each piece of equipment equals the width times the depth. This area times the expansion factor equals the floor area required by the equipment. These square-foot numbers are then added to determine the area required by the work center in the new layout, and the total appears at the end of the spreadsheet. This total area includes aisles, posts, and other routine intrusions upon the

```
                                      DISKETTE FILE: LIST.WK1
 A    B    C    D        E           F     G    H    I    J          K

 6 EXPANSION FACTORS:           3.8 = PRODUCTION AREAS
 7                              3.4 = STOREROOM, OTHER AREAS
 8                              4.1 = OFFICE AREAS, ETC.
 9                              5.8 = SHIP AND RECEIVE AREA
10 COLUMNS:        EQUIP. NUMBERS = NUMBER ASSIGNED TO EACH EQUIPMENT.
11                        FRM   TO = INCLUSIVE EQUIPMENT NUMBERS.
12                           TOT = TOTAL PIECES OF EQUIPMENT.
13                     W, D, H = WIDTH, DEPTH, HEIGHT (IN. OR FT.).
14                     SQ. FT. = EQUIPMENT AREA TIMES MULTIPLIER.
15                           * = NEW EQUIPMENT TO BE PURCHASED.
16
17                                                   UTILITIES
18                                                   ---------------
19   EQUIP.                                          CA = COMP. AIR
20 NUMBERS           WORK CENTER              SQ.    DI = DI WATER
21 FRM  TO  TOT    EQUIPMENT LIST     W    D    H    FT.   NI = NITROGEN
22 ---  ---  ---  ------------------  ---  ---  ---  ---   ---------------
23                     PRODUCTION
24                ---------------- 3.8
25 100  101   2  BENCH, TECH., HIGH   72   30   3'   114  110V, CA
26 103  108   6  BENCH, WORK, LOW     60   30   30   285  110V, CA
27 109  109   1  CABINET              36   15   6'    14
28 110  111   2  SHELF RACK           48   18   7'    46
29 114  114   1  TRASH CAN            24   24   2'    15
30 115       18  SHELF CART           36   18   5'   308
31 116  116   1  STAIR STEP           24   36   3'    23
32 117  117   1  SHELF RACK, HRDWR    48   18   7'    23
33 118  120   3  CABINET              36   15   6'    43
34 121  121   1  TRASH CAN            24   24   2'    15
35 122       10  LOCKER, DOUBLE     * 12   12   6'    38
36 123  123   1  RACK, BURN-IN       144   48   6'   182  110V 200A, CA
37 124  124   1  CABINET              36   15   6'    14
38 125  125   1  LAB CART           * 30   18   3'    14
39           1  CONVEYOR (FEET)     * 80   18   30   456
40
41                     SUPERVISOR
42                --------------- 4.1
43 200  200   1  DESK, STANDARD       60   30   30    51  110V
44 201  201   1  FILE, CABINET        15   24   4'    10
45 202  202   1  CHAIR, OFFICE        18   18   1'     9
46
47                OFFICE & ENGINEERING
48                --------------- 4.1
49 500  503   4  FILE, CABINET        15   24   4'    41
50 504  506   3  DESK, STANDARD       60   30   30   154  110V
51 507  509   3  CABINET, STORAGE     36   15   6'    46
52 510  510   1  TABLE, STANDARD      48   30   30    41  110V
53 511  512   2  FILE, CABINET        15   24   4'    21
54 513  513   1  DESK, STANDARD       60   30   30    51  110V
55 514        2  CHAIR, OFFICE      * 18   18   1'    18
56 515  515   1  TREE, OFFICE       * 24   24   4'    16
57 516  516   1  CHAIR, EASY        * 36   36   2'
58 517  517   1  TABLE, MAGAZINE    * 18   12   30

 A    B    C    D        E           F     G    H    I    J          K
```

Figure 5.1   New plant equipment list.

work center area. Measure all room dimensions from the center-line of the walls so the total area of all rooms will be close to the total area of the plant.

The IE arrived at these expansion factors from experience with other plant layout projects. The factors vary according to the equipment density characteristic of each type of work center or department. As you would expect, an office should have a more open ambience than the production floor, and that is reflected in the higher expansion factor assigned to it.

```
     A     B     C     D        E              F      G      H      I      J              K

           NUMBERS                WORK CENTER                            SQ.
           FRM   TO    TOT   EQUIPMENT LIST          W      D      H      FT.    UTILITIES
           ---   ---   ---   --------------  ---    ---    ---    ---    ---    ----------------
                             THICK FILM
     69                      --------------- 3.4
     70    701   701    1    BENCH, HOOD             60     30     6'     43     110V, CA, DI
     71    702   703    2    CABINET, STORAGE        36     15     6'     26
     72    704   704    1    FURNACE, CONVEYOR      144     36     4'    122     208V, 100KW, NI
     73    705   705    1    BENCH, TECH., HIGH      72     30     3'     51     110V, CA
     74    706   706    1    SCREEN PRINTER          30     36     4'     26     110V, CA
     75    707   707    1    OVEN                    24     24     5'     14     110V
     76    708   708    1    WALL HOOKS, SMOCK       36      6     5'      5
     77    709   709    1    TRASH CAN               24     24     2'     14
     78    710   710    1    FURNACE, CONVEYOR  *   132     36     4'    112     208V, 100KW, NI
     79    711   711    1    TRIMMER, LASER     *    60     24     5'     34     110V, CA
     80    712          2    LAB CART           *    30     18     1'     26
     81
     82                      SHOP
     83                      --------------- 3.4
     84    400   400    1    SHELF RACK              48     18     7'     20
     85    113   113    1    SHELF RACK              48     18     7'     20
     86    800   800    1    BENCH, TECH., HIGH      72     30     3'     51     110V, CA
     87    126   126    1    DRILL PRESS        *    12     18     5'      5     110V
     88
     89                      LUNCH ROOM
     90                      --------------- 4.1
     91    901   901    1    TABLE W/ CHAIRS    *    60     36     30     61
     92    902   903    2    VENDING MACHINE    *    30     24     6'     41     110V
     93    904          4    CHAIR, OFFICE      *    18     18     1'     37
     94
     95                      INSPECTION
     96                      --------------- 3.4
     97    401   401    1    SHELF RACK              48     18     7'     20
     98    403   403    1    BENCH, WORK             72     30     30     51     110V
     99    404   405    2    FILE, CABINET           15     24     4'     17
    100    801   801    1    CABINET, STORAGE        36     15     6'     13
    101    406   406    1    LAB CART           *    36     24     1'     20
    102                      CONVEYOR (FEET)    *    10     18     30     57
    103
    104                      OTHER ROOMS AND ITEMS
    105                      ---------------
    106                  2   REST ROOMS (SQ FT)                           250
    107                  1   UTILITY ROOM (SQ FT)                          90
    108     1     6     6    FIRE EXTINGUISHERS

     A     B     C     D        E              F      G      H      I      J              K
```

Figure 5.1 New plant equipment list (Continued).

These expansion factors are not chiseled in stone, and may have to be adjusted according to your industry or the needs of the personnel involved. To determine what expansion factors you are now using in a room, measure the areas of all the pieces of equipment. Add them up and divide the sum into the area of the room. The result is the area expansion factor for that room. If a more open effect is desired in the new plant, you can increase the factor slightly. Then multiply the factor by the sum of the footprints to arrive at the new area that will be required.

In most cases, the footprint is adequate for this calculation unless there is a considerable overhang. If that is the case, the footprint must still be recorded for use on the layout. This type of equipment will be treated in a special manner as you make the layout, as will be explained in later chapters.

A	B	C	D	E	F	G	H	I	J	K
	NUMBERS			WORK CENTER					SQ.	
	FRM	TO	TOT	EQUIPMENT LIST		W	D	H	FT.	UTILITIES
	---	---	---	------------------	---	---	---	---	---	----------------
				THICK FILM						
69				---------------- 3.4						
70	701	701	1	BENCH, HOOD		60	30	6'	43	110V, CA, DI
71	702	703	2	CABINET, STORAGE		36	15	6'	26	
72	704	704	1	FURNACE, CONVEYOR		144	36	4'	122	208V, 100KW, NI
73	705	705	1	BENCH, TECH., HIGH		72	30	3'	51	110V, CA
74	706	706	1	SCREEN PRINTER		30	36	4'	26	110V, CA
75	707	707	1	OVEN		24	24	5'	14	110V
76	708	708	1	WALL HOOKS, SMOCK		36	6	5'	5	
77	709	709	1	TRASH CAN		24	24	2'	14	
78	710	710	1	FURNACE, CONVEYOR	*	132	36	4'	112	208V, 100KW, NI
79	711	711	1	TRIMMER, LASER	*	60	24	5'	34	110V, CA
80	712		2	LAB CART	*	30	18	1'	26	
81										
82				SHOP						
83				---------------- 3.4						
84	400	400	1	SHELF RACK		48	18	7'	20	
85	113	113	1	SHELF RACK		48	18	7'	20	
86	800	800	1	BENCH, TECH., HIGH		72	30	3'	51	110V, CA
87	126	126	1	DRILL PRESS	*	12	18	5'	5	110V
88										
89				LUNCH ROOM						
90				---------------- 4.1						
91	901	901	1	TABLE W/ CHAIRS	*	60	36	30	61	
92	902	903	2	VENDING MACHINE	*	30	24	6'	41	110V
93	904		4	CHAIR, OFFICE	*	18	18	1'	37	
94										
95				INSPECTION						
96				---------------- 3.4						
97	401	401	1	SHELF RACK		48	18	7'	20	
98	403	403	1	BENCH, WORK		72	30	30	51	110V
99	404	405	2	FILE, CABINET		15	24	4'	17	
100	801	801	1	CABINET, STORAGE		36	15	6'	13	
101	406	406	1	LAB CART	*	36	24	1'	20	
102				CONVEYOR (FEET)	*	10	18	30	57	
103										
104				OTHER ROOMS AND ITEMS						
105				----------------						
106			2	REST ROOMS (SQ FT)					250	
107			1	UTILITY ROOM (SQ FT)					90	
108	1	6	6	FIRE EXTINGUISHERS						
A	B	C	D	E	F	G	H	I	J	K

Figure 5.1   New plant equipment list (Continued).

Notice the equipment numbers in the left-hand column of the spreadsheet. Each piece of equipment is assigned a separate identification number that will be used for three purposes:

1. In the check-off list to ensure that every piece of equipment has been identified and measured.

2. As an identifier, so people assigned to the area can easily recognize the equipment on the old plant layout and find it on the new plant layout during the layout review and revision phase.

3. As an equipment and location identifier on moving day so the equipment can be placed quickly when it arrives at the new plant. A color-coded, numbered sticker will be placed on each piece of equipment before it is moved to the new plant.

The formulas shown below are for the entries in the first cell of each range in Fig. 5.1, the new plant equipment list. The formula in the second cell of the range will be relative to the first cell. For instance, in the first item, the cell value for A6 is @CELL("ROW",A6..A6). The cell value for A7 will be @CELL("ROW",A7..A7), and so on.

Columns A to J = All columns are labeled at the top and bottom of each page.

A6 to A177 = @CELL("ROW",A6..A6)

This formula assigns numbers to the rows of the spreadsheet for demonstration purposes and will not be needed for your analysis.

B25 to B143 = First equipment number

IE entered data. This is the number assigned to the first piece of equipment in each category described in the EQUIPMENT LIST column.

C25 to C143 = Last equipment number

IE entered data. This is the number assigned to the last piece of equipment in each category described in the EQUIPMENT LIST column. If there is only one piece of equipment, then the first number is repeated, as in number 109.

D25 to D143 = C25-B25+1

This formula calculates the number of pieces of equipment represented by the numbers in columns B and C.

E25 to E144 = Work center equipment list

IE entered data. The IE inventoried all equipment to be located in the new plant and entered the description in this column.

F6 to F9 = Expansion factors

IE entered data. The area of each piece of equipment is multiplied by these factors, and the products are added to arrive at the estimated area required to house the equipment in a reasonable layout.

F24 to F139 = +F6

The expansion factor for each area is shown just below and to the right of each area title. These factors vary and refer to the figures shown at the top of the sheet in column F.

G10 to G15 = Column heading explanations

The column heading abbreviations are explained in more detail.

G25 to G144 = Width of each piece of equipment

IE entered data. The IE measured the width in inches across the front of each piece of equipment as you would face it when you use it, and entered the value in this column. Notice the conveyor width is given in feet.

Figure 5.2   Cell description.

H25 to H144 = Depth of each piece of equipment

IE entered data. The IE measured the depth in inches of each piece of equipment and entered the value in this column.

I25 to I144 = Height of each piece of equipment

IE entered data. The IE measured the height, to the nearest foot, of each piece of equipment and entered the value in this column.

J25 to J144 = @ROUND(F$24*D25*G25*H25/144,0)

This formula calculates the square feet allotted to each category of equipment, using the expansion factors assigned to each work center.

K25 to K144 = Utilities

The IE entered the utilities required for each category of equipment.

C151 to C160 = +F24

The expansion factors for each work center are repeated.

E151 to E160 = List of the work centers

J151 to J160 = @SUM(J24..J38)

The estimated area required for each work center.

J162 = @SUM(J150..J161)

The total area required for all work centers.

K151 to K160 = Actual work center areas

The actual areas assigned to each work center in the new plant layout of the Flexible Division.

K162 = @SUM(K150..K161)

The actual total area assigned to all work centers in the new plant layout of the Flexible Division.

E174 to E177 = Equipment to be released

This is a list of equipment that will not be needed in the new plant as a result of the new production plan and layout.

Figure 5.2 Cell description *(Continued).*

The From ("FRM") column is for the first number of the group and the TO column is for the last. These numbers can be used inclusively to cover several pieces of identical equipment. An example is the six workbenches, numbers 103 to 108. Some identical pieces of equipment can even be assigned the same number if they are mobile. An example is the 18 shelf carts, equipment number 115.

The Total ("TOT") column automatically calculates the number of common pieces of equipment by using the data that you supplied in the first two columns. After the description column, the

W, D, and H columns will contain your measurements for the width, depth, and height of each piece of equipment. The area of the equipment times the expansion factor is shown in the Square Feet ("SQ FT") column. The expansion factor for each work center is shown just below and to the right of the work center title. A more complete description of the spreadsheet formulas is given in Fig. 5.2.

The last column contains the utilities required for each type of equipment. This utility data can be entered while you are measuring the equipment or later, when you are planning the placement of the equipment in the layout.

Equipment Dimensions

Measuring the equipment is an onerous and unrewarding task, but one that is necessary for the layout, and it must be done accurately. A word of caution about allowing someone else to do the measuring for you: Don't let them do it. Do not assign this measuring work or any other measuring work to someone else. Inevitably, mistakes will be made by anyone who is not directly responsible for the accuracy of the end result. It is not that others are irresponsible or do not take the work seriously, but they generally cannot maintain an attitude of constant, meticulous accuracy.

Even when you do the work yourself, you must be uncompromising and alert to the possibility of errors. If you find yourself becoming bored or fatigued by the repetitive work, stop and resume the task later. Inaccurate measurements at this stage, or at any later stage, can cause problems on moving day that could compromise the smooth appearance and efficiency of the entire layout.

The measuring process can be performed with a minimum amount of time and effort by using the following procedure:

1. Make a sketch of the walls of the area or room to be measured, and include a sketch of each piece of equipment. Just use rectangles for most of them, leaving space next to each piece so you can record the dimensions.

2. If some equipment is more complicated, requiring several rectangles, then sketch those also.

3. Some equipment may have other rectangles on top of the main body that would appear on the 3-D view later. Sketch those in also. Keep it simple, and show just enough detail to transmit the impression of the equipment, not a picture of it.

4. Measure the equipment, and write down the dimensions of each piece next to it on the drawing. The width of the equipment is measured across the part that you face when using the equipment. The depth is measured straight back, and the height is from the floor to the top of the uppermost member.

5. If you plan to use a CAD program to make the layout drawing, then record all your dimensions to the nearest 3 in. The

SNAP distance on the screen will be set to 3 in also. That is, all dimensions on the final layout will be to the nearest 3 in, including the equipment, walls, doorways, or any other structure that appears on the layout. Only in very rare cases will more accuracy be required.

6. Save all the data that you accumulate for the layout in a notebook. The equipment sketches should definitely be saved, to avoid a return visit just to remeasure something. The notebook will be completely full of sketches and other data by the end of the project.

When you have measured all the equipment for one area, you can enter the measurement data on the spreadsheet. Just load your Lotus or other spreadsheet program and retrieve the LIST.WK1 file. You can change the department headings, the equipment descriptions, and the dimensions. The rest of the data will be calculated automatically from your entries. The summary at the bottom of the sheet will calculate the square feet required for each department as well as the total area of the plant. Even the department headings will be transferred automatically to the data summary section.

If you want to change the value of the expansion factors, do so in the upper, constant-value section. They will then be transferred automatically to each area factor in the center section. The expansion factors in the center section have been assigned for particular types of work centers, so try to use the same type of heading for each section. For example, the supervisor work center has an office area expansion factor assigned to it. If you change the heading, try to change it to an office type of work center, or you will have to change the factor also. After you are through, check the factors for each of your headings, and change them if necessary.

Bear in mind that the expansion factors shown at the top of the spreadsheet, used to calculate the areas of each work center, are given only as a rough guide. The expansion factors can be used to evaluate the space available in a plant under consideration, but should not be the sole determining factor. If you have any doubt about whether the new plant floor space is adequate, make a rough sketch of a possible layout of the major pieces of equipment.

Having the right amount of floor space is not always enough either. Even if the new plant area is adequate, it may not suit your application because the walls or columns are not located properly. This is especially true of a warehouse, where pallet racks or shelving must be located at fixed intervals, and you must avoid having a column right in the middle of an aisle.

If that is your situation, make a brief layout of your pallet racks or shelf racks to see how they fit between the columns of the new plant. If they do not fit conveniently, you could lose a sizable amount of floor space when you have to compromise the layout to fit the column spacing.

Other Requirements

Although the floor space available is one of the most important considerations in the choice of a new plant, there are many other items to consider also. If you are having the new plant built, the architect will automatically incorporate many of them. If you are buying or leasing the new plant, consider these items before making your choice:

1. *Room for expansion.* This is possibly the most important item. Before you recommend a new building, ask all the managers if they have any plans for future expansion. If they do, then allow space for it.

2. *Electric power.* The total electrical requirements may have to be calculated to ensure that adequate power is available from the local utility company. This normally will not be a problem unless your plant contains high-powered equipment such as environmental chambers, conveyor furnaces, batch ovens, power transformers, many welders or other special heating requirements, etc.

3. *Special gases.* These gases might include carbon dioxide, oxygen, hydrogen, helium, nitrogen (gas or liquid), argon, etc. There must be room outside for the supply containers and access for the delivery trucks. Also, if the gases are going to be piped into the plant through an expensive distribution system, the distance between the supply tank and the user should be as short as possible. An example would be cryogenic piping for liquid nitrogen.

4. *Heating, ventilating, and air conditioning (HVAC).* If you have production equipment that radiates heat into the plant, the HVAC system may have to be modified to compensate for the additional load. The extra heat could raise the ambient temperature above an acceptable level, and it may be too late to change the air conditioning system after you have moved in. The best insurance against such a calamity is to have an HVAC engineer survey your heating and cooling requirements and make recommendations for a solution before you decide on which plant to buy or lease.

5. *Rest rooms.* There may be adequate rest rooms in the plant already, but a brief survey is still advisable. Some initial guidelines are shown in the plant layout guidelines of Fig. 12.4, but call your local building code inspector if there is any doubt about their adequacy.

6. *Shipping and receiving.* The number of bays should be adequate, and they should be at the right height for the delivery trucks. If the bays are not sealed when the trucks are in place, will the prevailing wind cause a problem? Is there enough floor space for both shipping and receiving so they can be properly divided? Can the inspection and storeroom areas be located nearby for easy transfer of material to them? There should be a separate door to the outside besides the roll-up doors. Employees should not have to use this area to enter the plant. Large delivery

trucks should have room to maneuver and back up into the bay easily. There should be adequate heating facilities for this area in the winter even when the traffic is heavy.

7. *Miscellaneous.* This category includes all those requirements that may be unique to your operation. Examples include floor drains, employee access doors, lunch room, cafeteria, conference rooms, break areas, reception area, computer control room, telephone room, utility room, drop ceiling, production area lighting, outside windows, wide doors, adequate insulation, and many more. Make your own list as you tour the new building and contemplate what is in the old plant that is necessary for your operation.

8. *People with disabilities.* There are many requirements for a plant to accommodate people with disabilities, more than can be listed in this book. Only new construction must comply with the Americans with Disabilities Act of 1990. Access ramps, rest rooms, and workplace enhancements are just a few of the accommodations that must be considered. This subject will be covered in more detail later in the book. For now, you can use *The Americans with Disabilities Act Checklist for Readily Achievable Barrier Removal*,[19] to check out the new plant for ADA compliance. Make a survey of your present and future needs, and decide if any extra work will be required for the new plant.

9. *Parking lot capacity.* Because it is not involved in the layout, the parking lot is often overlooked when new plant alternatives are being considered. How many plants have you seen that have parking lots too small for the number of employees? The streets around the plant are packed with cars, and local business owners are furious. Before you make the final decision on buying or leasing a plant, be sure that the parking lot is large enough for the present number of employees and for future expansion.

Work Center Relationships

The subject of work center relationships has been studied to exhaustion. Many computer programs have been published for the sole purpose of developing spacial diagrams of the optimum location of related work centers. In the manufacturing arena, most departmental functions and their working relationships are so well known that further study is not necessary. If your plant operation is more complex or if you are not familiar with the standard manufacturing routine, ask the key people in your organization to describe the function of their work centers. If you find that more analysis is necessary, refer to the *Handbook of Industrial Engineering*,[1] page 1797, where the development of a complex interrelationship diagram is discussed.

The simple relationship diagram shown in Fig. 5.3 is typical of the core of a high-tech manufacturing system, and is adequate for laying out most production activities. This drawing is on the diskette as an AutoCAD file named RELATE.DWG. This type of analysis, called a *bubble chart* or *block diagram*, has been in use for decades. In fact, it was very likely used in some form by Henry Ford to develop the Model T moving assembly line in 1913.

You may have to lay out other activities also, such as marketing, sales, accounting, human resources, research and development, documentation, drafting, production control, and materials. If they are to be included in the relationship diagram, then add ellipses for each activity and indicate the degree of association with other activities. The number of solid lines between the bubbles in Fig. 5.3 indicates the relative volume of material flow. Again, most information conveyed by the diagram will be obvious to anyone familiar with the plant operation, and it is not cost-effective to spend a great deal of time documenting it.

This type of relationship diagram has one use that is not often recognized: It can be used as a check for errors after the work

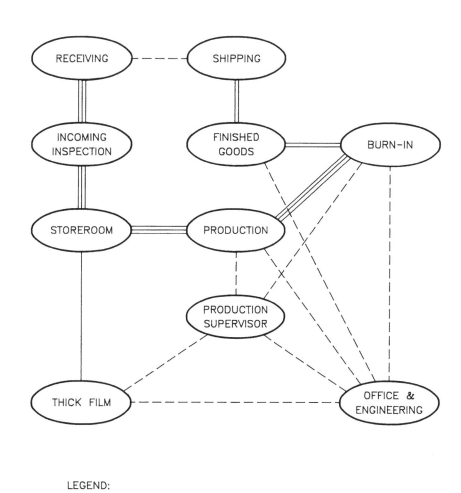

Figure 5.3 Relationship diagram.

center locations have been assigned. That is, most work centers are assigned a location in the new plant by using the obvious relationship or an available room. If a relationship diagram is then made showing the assigned placement of these work centers and their degree of connection, it may disclose some obscure error that could affect the overall layout. In the end, you should judge whether you will need a relationship diagram or not.

The relationships shown in Fig. 5.3 conform to the actual work center arrangement that will be used in the Flexible Division new plant layout. The IE prepared this diagram to verify that there were no obvious errors and as part of the package presented to management to support the choice of a new plant. The package also included the equipment list, an advantages list, and a disadvantages list.

This type of relationship diagram can be used for other facilities also. An order-picking operation can be described in the same terms as a manufacturing plant. The only real difference is the way the product is managed. In a manufacturing plant, the parts are unpacked, some value is added, and then they are assembled into a product and packaged for shipment. In an order-picking operation, the parts are unpacked, repacked according to a pick list, and prepared for shipment.

At this point in the project, you should be able to decide if a particular building will be suitable for your company. By combining all the separate analyses that have been presented so far, you can propose an improved production plan and qualify the building which will house it.

6

Project
Organization

This chapter describes some important problems connected with preparing a plant layout project. Also, the involvement of the general contractor will be explained, and some examples of high-priority considerations will be presented. The Flexible Division project schedule will be illustrated.

Purpose of a Plant Layout

Unless you are a new employee, you are probably very familiar with the production system in your plant. If you have conducted the analysis presented in the first few chapters of this book, you have an excellent concept of how the old system operates and how to improve its efficiency. But before you start, consider some of the purposes of a plant layout:

1. To plan where the equipment will be placed.

2. To plan the location of the various work centers.

3. To plan for adding new equipment.

4. To plan for discarding old equipment no longer needed.

5. To allocate the new floor space.

6. To plan the utilities.

7. To conclude a project to improve the manufacturing process.

The last item has the most potential for realizing the benefits of a new plant layout. If the plant move can be anticipated by months or years, a plan can be developed to improve the product quality, the flow, the product cost, and the overall output of the entire system. The new plant layout will play a key role in accomplishing the benefits of this plan. If the plant layout project is fully developed, the effect on the company's future can be dramatic.

Preliminary Decisions

When a company builds a new plant, the layout designer seldom participates in the actual design of the building, except for supplying the floor space data and the work center relationships. Normally, the architect is given this design data, and some additional guidelines are supplied by management. The end result is a building that approximates the desired layout goals and will usually satisfy the needs of the manufacturing system. During this phase, more emphasis is placed on the exterior appearance, the reception area, and the managerial office complex.

Instead of building a new plant, however, it is more common for a company to move to an existing building and modify it to suit the needs of its manufacturing system. That is what the Flexible Division is doing. It is moving to an existing building next door to the corporate headquarters. For that reason, the Flexible Division and the parent company can share the existing office organization that is already in place.

One of the primary decisions at this stage is whether the new building will satisfy the layout goals. The new plant may or may not contain all the features that are needed to perform the manufacturing function of your company. In Chapter 5, a detailed method of qualifying a building was presented, but that degree of refinement is not always required, or the time may not be available.

If time is short, the best way to qualify the area of a new plant is to design a rough layout for it. Or you can measure the square-foot areas of the existing layout, add in the anticipated areas for new equipment or functions, and balance these figures against the area of the new plant. At this point, accurate measurements of the buildings are not necessary. The areas of the new and old plants can be paced off and tabulated rather quickly.

If the result is a close fit, you may want to make more exact measurements and calculations. It may still be a judgment call as to whether the new plant will be adequate. Generally, if it appears to be a close fit, then the new plant is probably too small. There will be numerous demands made on the new building floor space that cannot be anticipated this early in the undertaking. Once you commit to a new building, those demands will forever plague the layout project. The final decision will be based on your impression of how crowded the present plant is, what the goals of the company are, and how the new plant will satisfy those goals.

A word should be mentioned here about how much time will be needed for planning a plant layout. The actual project can be planned and executed in a very short time if that is required, but there is a price to pay, and it can be extremely high. A plant layout project encompasses so many facets of the company's operation that if it is rushed, time will be lost in the end, not gained. Adjustments to the layout will be inevitable and the project will cost more in time and money than if the layout had been planned right in the first place.

The important point to remember is this: A plant layout can always be designed in the time that is available, even overnight if necessary. The benefit to the company, however, will always be commensurate with the time devoted to the layout planning phase. In the plant layout project schedule of Fig. 6.1, the Flexible Division took six months to complete the layout planning phase and over a year to complete the project. The result, however, elevated the company to a position of leadership in its industry.

Assuming that you have been allowed sufficient time to perform the project and that the building has been chosen, tour the new plant at the first opportunity. The object of this first examination is to appraise the restrictions that the new plant will impose on the layout and to observe the available utilities. The structure of the existing new plant will also be important. During the tour, notice these items:

1. *Columns.* They will form the center points for all measurements on the layout. Are they in line, and do they appear to be a uniform distance from each other and from the walls?

2. *Power panels, floor drains, water supplies, etc.* Will they subtract a substantial amount of floor area from the layout? If they are needed, are they located where they can be used?

3. *Overhead.* Will a drop ceiling be required to create the desired ambience for your working environment? How will the maintenance group be able to service the utilities above this ceiling?

4. *Doors.* Do the doors vary in size, and are they large enough for your equipment? Will they have to be replaced, or are they all at least 3 ft wide? Are there double doors leading to the production area from the receiving and shipping areas?

5. *Windows.* Are there any internal rooms that will need windows to create a more open environment for the workers inside?

6. *Repairs.* Will any important repairs have to be made immediately before other work can begin? These items might include a faulty roof, peeling walls, inoperative furnace, broken outside doors, leaking water main, etc.

7. *Dangerous conditions.* Is asbestos insulation present in the walls or around any of the ducts or pipes? Is gas leaking inside or outside the building? Is part of the building framework in need of reconstruction? Do chemicals have to be removed? Has the ground under the floor drains and outside dumping areas been tested, and is it free of contamination?

8. *Office area.* Is the office area generally adequate or expandable? Is there enough room for the managers, the reception area, and other special uses such as sales conference rooms?

9. *Other special requirements.* Do any of your operations have special requirements, such as clean rooms, plating rooms, special gases, or inspection rooms? Will you need a remote storage building for flammable materials?

After the tour of the new plant is complete, conduct a detailed inspection of your old manufacturing areas, utilities, and support group functions. This may seem like a backward approach to tour the new plant first, but it will be easier to remember the new plant while looking at the old one than the other way around.

While you are reviewing the old plant production system, you can mentally place many of the work centers in the new building simply because of their special requirements. For example, the plating room will need that area in the new plant with good floor drainage. Or the mask aligning room will need that area in the new plant with an isolated floor slab.

If the old or new plant has complicated features, take an instant camera along. It is surprising how just a few photos can save you many trips back to an area to verify your recollection. In some cases, you can even make rough measurements right from the pictures. In general, this initial review will start the thought process that will help you plan the entire project, including the final move.

General Contractor

If you are moving to an existing plant, you will usually have to make some structural and utility modifications. These changes will normally be made by a general contractor. In the case of a new building, the architect will be involved, and the initial planning will start well in advance of construction. If you are assigned to the project early, it will allow you more time to work with the architect and customize the new plant exactly to your layout needs. Work with the architect right from the start if that is possible. But, regardless of the sequence of events, establish a close association with the general contractor during the entire project.

General contractors operate their enterprises under different rules than almost any other business. Their objectives appear to be similar, but they accomplish them differently. These objectives are:

1. To satisfy the customer and make a profit.

2. To perform the desired work, including the subcontract work, as the contract was written.

3. To keep their own workers employed on a full-time basis.

4. To complete the project within a reasonable time.

5. To recognize that additional work may be required for the project and ensure that it is profitable.

In order to relate to the contractor effectively, you must sympathize with the profusion of problems that can occur during construction. The contractor may have several building jobs running sequentially with numerous reliable and unreliable subcontractors. At one time or another, they will all offer excuses for their degree of nonperformance. The jobs must be scheduled to overlap slightly to keep the contractor's employees busy, but if one job runs late, the customer will be dissatisfied. It is a constant juggling act and one of the most thankless occupations in the manufacturing industry.

Establish a sympathetic and friendly relationship with the general contractor from the start. Your company may be paying for the work, but the contractor is orchestrating the renovation. That is just another anomaly of the contracting business. If you have a serious disagreement which cannot be resolved, refer the problem to management.

If possible, become involved in specifying the work described on the original contract, and make sure that all renovations are included. It is important to avoid adding extra work after the contract has been accepted; the contractor must spend more time planning this extra work, so the profit must be higher. The general contractor would be in contractor's heaven if it were possible to plan on doing what is in the contract and no more.

If there are changes in any construction details after the work has started—adding doors, for instance—present them in one organized package. Don't hand the contractor a series of rough notes on slips of paper. Do present an organized list of the work to be done, described in enough detail so it cannot be misinterpreted. If you think there could be some doubt about what you mean, have the contractor explain each item back to you. The contractor will appreciate this extra effort as much as you will.

This added-work situation can be the major cause of animosity between you and the general contractor: The work may be in error, you become agitated about it, the contractor gets angry with you for changing your mind, and you are both at a standoff, creating a very hostile relationship. Avoid this situation at all costs. The general contractor can make or break the fine-tuned, finished appearance of your plant, and must be appreciated for that fact.

Much of your success as renovation director depends on your relationship with the general contractor. An extra convenience, such as a pass-thru in a wall between work centers, may cost an exorbitant amount of money if it is installed at all, or it may be done at a reasonable cost as a gesture of friendship.

There is one more negative aspect to this added-work situation. When the general contractor postpones the completion date of the contract, which is almost inevitable, the contractor will state that the added work caused the delay. And if it was sizable, that will be true. This situation serves to further emphasize the

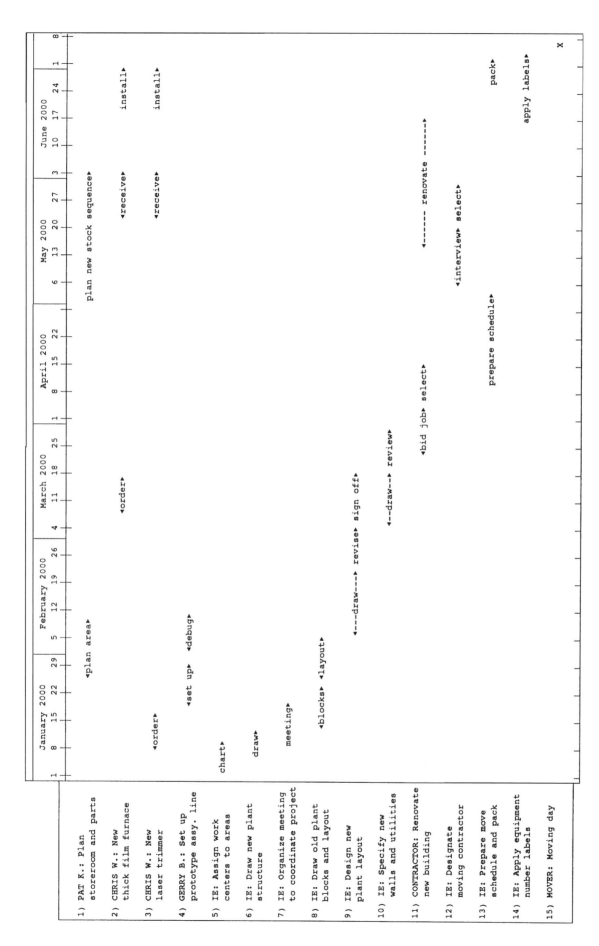

Figure 6.1 Plant layout project schedule.

importance of specifying the total job on the original contract and avoiding changes as much as possible.

In later chapters, this subject of anticipating and planning all the renovation work will be expanded. The construction and utility specification drawings made by the IE to avoid these problems will be described and illustrated.

Project Schedule

An example of a plant layout project schedule is shown in Fig. 6.1. It is a typical Gantt chart, which is time-based, and has many advantages over a Pert chart, which is task-based. To schedule an uncomplicated layout project like this, simple charts are better. The purposes of the schedule are as follows:

1. To organize the project into logical, sequential tasks.

2. To assign responsibility for those tasks.

3. To provide a time-based guide for the task leaders.

4. To provide a progress guide to management.

5. To confirm that the project can be finished on time.

6. To support management's confidence in the project's success.

7. To provide a record of actual completion dates after the project is finished.

The last item is an added benefit that is not often realized by the project leader. If you record the actual start and completion dates of each task, the schedule will become a valuable tool for scheduling future projects. It can also be useful when you have to field questions from management about the actual completion dates of some tasks.

The initial schedule must, of necessity, be very preliminary and will be updated often as the project progresses. You can list all the tasks with some accuracy, but you cannot schedule exact completion dates at this time. It is important, however, to have a plan of action right from the start. Even if the dates are only estimated, the other task leaders and management will appreciate the advance warning of what is coming and a reasonable estimate of when it will happen.

The schedule shown in Fig. 6.1 was constructed using the Quick Schedule Plus program.[8] This inexpensive program is very easy to learn and use. If you need a schedule program for brief projects once or twice a year, it is ideal. The schedule shown in Fig. 6.1 is also on the diskette as a file named SCHEDULE.QS.

The *heading* is the title of the task and appears in the left-hand column. The *bars* are separate parts of the task and appear to the right of the heading. For instance, for task 2, *CHRIS W.: New thick-film furnace*, is the heading, and "<order>" is a bar for that task.

Quick Schedule Plus is one of the few programs that allow you to enter descriptive text right into the task bars. Notice that task 3 has three bars: "<order>," "<receive>," and "install>." All are described on one line, and because of this capability, the whole schedule can be shown on one sheet. Less convenient programs would require three lines to schedule this same task and could require two sheets for the entire schedule.

The bars of a single task can be moved horizontally, one at a time or as a group. The bracket symbols for each bar can be changed. The time intervals at the top of the schedule can be changed to days, weeks, or months. Notes can be attached to a task and shown on the printout. Numerical data, worker hours, for instance, can be summarized at the bottom of the schedule.

There are some disadvantages to Quick Schedule Plus. Several tasks cannot be moved horizontally as a group. The heading and task bars cannot be moved vertically as a unit. The bars can be moved, but the task must be retyped. The schedule can only be printed by a printer that has a parallel connection. Even with these shortcomings, a brief schedule like this one can be constructed rather quickly with some advance planning of the task sequence.

There are several important details to recognize when you organize the schedule:

1. Number the tasks and assign responsibility to a task leader by name. Show both items in the heading.

2. Group the tasks in chronological order for each leader.

3. Allow a generous amount of time to complete a task if the leader has to add that work to an already full schedule.

4. Describe the task as clearly as possible in the heading.

In the Quick Schedule Plus presentation shown here, the start date is the symbol <, and the completion date is the symbol >. Sometimes the description within each bar is wider than the "time" space that is needed for that part of the task. In those cases, only the completion date is important, so the start date symbol is omitted. For example, "plan new stock sequence>" in task 1 will take only one week to accomplish, but the bar description encompasses five weeks, so there is no start date symbol. That is true for short jobs also; only the completion date is important, and the start date can vary. If a part of the task requires more time than the description uses, the extra space can be filled with dashes, as in "<---draw--->" in task 9.

The schedule shown in Fig. 6.1 will be used by the Flexible Division for its new plant layout project. Each of the tasks listed in the left column of the schedule will be mentioned later in the book, except for task 5, as the relationship drawing has already started. The IE is also the task leader for developing the improvements in the assembly procedure. Notice the timing of the various tasks assigned to the IE. Many of the bars overlap, but the completion dates form a steplike pattern that allows a sufficient time interval to complete each task.

The timing of the tasks in this schedule is based on laying out a plant of about 4000 sq ft by a project leader who can devote full time to the project. If your new plant is larger, or if you cannot devote full time to it, you will have to make appropriate adjustments in your schedule.

Notice that the total project consumes more than six months, and the time allotted for each task is adequate but not excessive. If you do not have time to complete all the tasks on your schedule, decide now what must be eliminated.

If the project time must be drastically reduced, it will be in your best interest to notify management of the consequences, by written memo. State where you will have to cut corners or what tasks will have to be eliminated in order to complete the most important tasks properly. If you are still pressured to do everything in a restricted time frame, mistakes will be inevitable, and more than likely, you will be blamed for them. Management will have more sympathy for your plight if that memo is in their files.

Here is a brief description of the tasks shown in Fig. 6.1 for the Flexible Division:

1. *PAT K.: Plan storeroom and parts.* The storeroom supervisor will design a new arrangement for the receiving, inspection, and storeroom areas. This includes specifying the reduced number of shelf racks that will be needed in the storeroom. Pat will also develop a plan to store the parts in kit order instead of stock number order on the shelves.

2. *CHRIS W.: New thick-film furnace.* The manufacturing engineer will specify and order a new thick-film conveyor furnace.

3. *CHRIS W.: New laser trimmer.* The manufacturing engineer will order the new laser trimmer. Notice the lead time that is needed for this equipment. Chris has to order it right away, before the new layout is even started.

4. *GERRY B.: Set up prototype assy. line.* The production supervisor will set up an assembly line to test the new assembly procedure. Gerry will not use a conveyor, but will have the operators pass the units manually instead.

5. *IE: Assign work centers to areas.* The IE will develop a relationship diagram that duplicates the work center arrangement in the new plant.

6. *IE: Draw new plant structure.* The IE will draw the structure of the new plant, assign work center areas, and indicate the material flow pattern.

7. *IE: Organize meeting to coordinate project.* The IE will call a meeting with Pat, Chris, and Gerry to discuss and approve the changes in the production system.

8. *IE: Draw old plant blocks and layout.* The IE will make a simple CAD drawing of the old plant, assign numbers to all equipment, and make blocks of the custom equipment that will be needed for the new layout.

9. *IE: Design new plant layout.* The IE will draw the new plant layout, review it with the users, revise it several times, and have it signed off by management.

10. *IE: Specify new walls and utilities.* The IE will make utility drawings, review them with Chris and Gerry, and have the drawings signed off by management.

11. *CONTRACTOR: Renovate new building.* The IE will request bids for the work, management will award a contract, and the IE will supervise the renovation.

12. *IE: Designate moving contractor.* The IE will interview the contractors and select the best one for performing the move.

13. *IE: Prepare move schedule and pack.* The IE will prepare and disseminate the move schedule. The IE will also be available for questions while everything is being packed or prepared for the move.

14. *IE: Apply equipment number labels.* The IE will apply number labels (color-coded by work center) in a prominent location on each piece of equipment.

15. *MOVER: Moving day.* The IE will ensure that the key personnel involved with the move are in place and that all problems are solved.

After your schedule is complete, you will be in an excellent position to advise management about what can and cannot be accomplished in the time frame that has been allotted to the plant layout project. This is the time to meet briefly with management, discuss the schedule, and reach an agreement on the scope of the project.

High-Priority Considerations

Most equipment that must be purchased can be delivered within one or two months, and can be entered in the layout schedule with a minimum of pre-planning. However, some pieces may have such a long lead time that it threatens to postpone the moving day unless it is arranged for immediately. These long-lead-time items, as well as any installations that require advance planning, must be resolved at the beginning of the layout project. Here are some examples:

1. *Cryogenic piping.* If low-temperature liquefied gas or similar uncommon supplies are used in your process, the distribution system of piping can be very expensive, and may have to be ordered six to nine months before delivery.

2. *Customized equipment.* If new equipment must be modified or constructed just for your process, it can take many months to prepare it for your plant.

3. *Installation priorities.* Some equipment, such as a hooded bench, requires special installation work. If your new plant is still under construction, the contractor may notify you that the sheet-metal worker is overbooked and will be in the plant for only two weeks, starting tomorrow. The contractor wants to know where the bench hood will be located. The roof vent and exhaust blower are on hand, but you will have to drop what you are doing and lay out that area right now.

4. *Electric power.* Some pieces of equipment have very high power requirements which should be considered early in the lay-out schedule. The contractor, the electrical subcontractor, and the utility company should all be notified of this situation as soon as possible. The high-powered equipment may have to be located on the layout right now, so plans can be made to route the power to the building and equipment at the lowest possible cost.

The above examples represent a host of problems that should receive immediate attention before you become engrossed in the plant layout. If you find problems that will delay the layout project and cannot be solved, notify management immediately. Always keep management informed of the status of the project. If there are no major problems, tell them that, too.

Plant Layout Blocks

This chapter will concentrate on designing blocks within the AutoCAD program. CAD users must take this preliminary step before designing the plant layout. Instructions will be provided for:

1. Using the library of equipment blocks and others that are supplied on the diskette.

2. Constructing your own blocks that are unique to your plant.

3. Recording the new blocks on the list of blocks.

Equipment Blocks

All the equipment blocks supplied on the diskette are three-dimensional, so when they are displayed on the screen by using the VPOINT command, or when a 3-D view is plotted, the equipment blocks will appear as three-dimensional images.

The entire library of blocks is listed in Fig. 7.1. This list is also supplied on the diskette in a WordPerfect file named BLOCKS.WP. As you make more blocks, you can add them to this list, which can be extended for many more pages. The columnar configuration of this file is not common, but is very useful for this purpose. As you add a block into any column, the program will automatically move the last entry in each column to the next column or onto the next page, as appropriate. Hard returns can be added or deleted to make sure each category remains undivided, as shown in Fig. 7.1.

It is important to understand the method of assigning file-names to these blocks, as it can also be applied to the new

BENCH, HOODED
```
6' X 36 X 6' = BH072366
8' X 36 X 6' = BH096366
```

BENCH, PLAIN
```
4' X 30 X 30 = BP04830-
5' X 30 X 30 = BP06030-
5' X 36 X 30 = BP06036-
6' X 30 X 30 = BP07230-
6' X 30 X 3' = BP072303
6' X 36 X 30 = BP07236-
6' X 36 X 3' = BP072363
8' X 30 X 30 = BP09630-
8' X 30 X 3' = BP096303
8' X 36 X 30 = BP09636-
8' X 36 X 3' = BP096363
```

BENCH, TECHNICIAN
```
6' X 30 X 30 = BT07230-
6' X 30 X 3' = BT072303
6' X 36 X 30 = BT07236-
6' X 36 X 3' = BT072363
8' X 30 X 30 = BT09630-
8' X 30 X 3' = BT096303
8' X 36 X 30 = BT09636-
8' X 36 X 3' = BT096363
```

BOOKCASE
```
3' X 12 X 3' = BC036123
3' X 12 X 6' = BC036126
```

CABINET
```
3' X 15 X 6' = CS036156
3' X 18 X 6' = CS036186
3' X 21 X 6' = CS036216
3' X 24 X 6' = CS036246
```

CHAIR, WORK
```
LOW       = CW018181
HIGH      = CW018182
EASY      = CW036362
```

COAT HANGERS
```
3' X 18 X 5' = CH036185
4' X 18 X 5' = CH048185
```

COLUMN, ROUND
```
 6 DIA. X 9' = CN006069
 8 DIA. X 9' = CN008089
10 DIA. X 9' = CN010109
12 DIA. X 9' = CN012129
```

COLUMN, SQUARE
```
 6 X  6 X 9' = CQ006069
 8 X  8 X 9' = CQ008089
10 X 10 X 9  = CQ010109
12 X 12 X 9' = CQ012129
```

CONVEYOR, CORNER
```
18 X 48 X 30 = CC01848-
2' X 48 X 30 = CC02448-
3' X 84 X 30 = CC03684-
```

CONVEYOR, FLOOR
```
20'X 36 X  6 = CF240360
20'X 48 X  6 = CF240480
```

CONVEYOR, GRAVITY
```
20'X 18 X 30 = CG24018-
20'X 24 X 30 = CG24024-
20'X 36 X 30 = CG24036-
```

CONVEYOR, TRANSFER
```
18 X 18 X 30 = CT01818-
2' X 24 X 30 = CT02424-
3' X 36 X 30 = CT03636-
```

DESK, FOREMAN
```
36 X 30 X 4' = DF036304
```

DESK, STANDARD
```
4' X 30 X 30 = DS04830-
5' X 30 X 30 = DS06030-
5' X 36 X 30 = DS06036-
6' X 30 X 30 = DS07230-
6' X 36 X 30 = DS07236-
```

DOOR, RIGHT
```
24" WIDE    = DR24
30" WIDE    = DR30
36" WIDE    = DR36
42" WIDE    = DR42
48" WIDE    = DR48
```

DOOR, LEFT
```
24" WIDE    = DL24
30" WIDE    = DL30
36" WIDE    = DL36
42" WIDE    = DL42
48" WIDE    = DL48
```

ESD SYMBOL
```
BENCH LOW   = ESDLOW
BENCH HIGH  = ESDHIGH
```

FILE CABINET
```
15 X 24 X 30 = FC01524-
15 X 24 X 4' = FC015244
15 X 27 X 30 = FC01527-
15 X 27 X 4' = FC015274
18 X 30 X 30 = FC01830-
18 X 30 X 4' = FC018304
```

FILE, LATERAL
```
30 X 21 X 4' = FL030214
36 X 21 X 4' = FL036214
42 X 21 X 4' = FL042214
```

FIRE EXTINGUISHER
```
12 DIA DRY   = FED12120
12 DIA CO2   = FEC12120
12 DIA HALON = FEH12120
12 DIA WATER = FEW12120
```

FLOW ARROWS
```
STRAIGHT    = FLOW-ST
RIGHT ANGLE = FLOW-RA
LEFT ANGLE  = FLOW-LA
RIGHT SLANT = FLOW-RS
LEFT SLANT  = FLOW-LS
```

FORKLIFT
```
10'X 36 X 5' = FK102365
```

LAB CART
```
30 X 18 X 3' = LC030183
3' X 24 X 3' = LC036243
```

LOCKER, SINGLE
```
12 X 12 X 5' = LK012125
12 X 15 X 5' = LK012155
12 X 18 X 5' = LK012185
```

LOCKER, DOUBLE
```
12 X 12 X 6' = LK012126
12 X 15 X 6' = LK012156
12 X 18 X 6' = LK012186
```

Figure 7.1   List of blocks.

blocks that you will design. If you use this system, it will make all your layout block filenames uniform and easily identified should they become mixed in with your other block files.

Notice the filenames of the blocks that are shown under each category. For instance, under BENCH, HOODED there are two sizes and two names. The first size is 6 ft × 36 in × 6 ft, and the filename is BH072366. The filename not only identifies the object, but also tells what the dimensions are, and it does all this by using only eight characters. The name can be divided into four parts:

1. BH = bench, hooded.

2. 072 = the width of the bench in inches.

```
MACHINE TOOLS
  DRILL PRESS   = MTDRILL
  GRIND WHEEL   = MTGRIND
  LATHE         = MTLATHE
  MILL          = MTMILL
  SURF. GRIND.  = MTSURF
  PUNCH PRESS   = MTPUNCH

MISCELLANEOUS
  DIMEN. ARROW  = MI-DIMAR
  DIVIDE BLK.   = MI-DIV
  FENCE         = MI-FENCE
  NORTH ARROW   = MI-NORTH
  TREE, OFFICE  = MI-TREE
  UTIL. ARROW   = MI-UTIL
  UTILITY POST  = MI-DROP
  DRINK FOUNT.  = MI-DRINK

PALLET RACK
  8'  X 36 X 8' = PR096368
  8'  X 42 X 8' = PR096428
  8'  X 48 X 8' = PR096488
  9'  X 36 X 8' = PR108368
  9'  X 42 X 8' = PR108428
  9'  X 48 X 8' = PR108488
  10' X 36 X 8' = PR120368
  10' X 42 X 8' = PR120428
  10' X 48 X 8' = PR120488
  11' X 36 X 8' = PR132368
  11' X 42 X 8' = PR132428
  11' X 48 X 8' = PR132488
  12' X 36 X 8' = PR144368
  12' X 42 X 8' = PR144428
  12' X 4' X 8' = PR144488

PALLET STACK
  3'  X 3'  X 5' = PS036365
  42  X 3'  X 5' = PS042365
  42  X 42  X 5' = PS042425
  4'  X 3'  X 5' = PS048365
  4'  X 42  X 5' = PS048425
  4'  X 4'  X 5' = PS048485

PALLET, PLAIN
  3'  X 3'  X 6" = PP036360
  42  X 3'  X 6" = PP042360
  42  X 42  X 6" = PP042420
  4'  X 36  X 6" = PP048360
  4'  X 42  X 6" = PP048420
  4'  X 48  X 6" = PP048480
```

```
PARTITIONS, OFFICE
  POST          = PT002025
  1' X  2 X 5'  = PT012025
  2' X  2 X 5'  = PT024025
  30 X  2 X 5'  = PT030025
  3' X  2 X 5'  = PT036025
  4' X  2 X 5'  = PT048025
  5' X  2 X 5'  = PT060025
  6' X  2 X 5'  = PT072025

PERSON
  FEM, STAND    = PEFESTAN
  FEM, SIT LO   = PEFESITL
  FEM, SIT HI   = PEFESITH
  MAN, STAND    = PEMNSTAN
  MAN, SIT LO   = PEMNSITL
  MAN, SIT HI   = PEMNSITH

PRINTER
  24 X 21 X 30  = PN02421-
  30 X 24 X 30  = PN03024-

RACK, CABLE
  24 X 24 X 4'  = RC024244
  24 X 36 X 6'  = RC024366
  36 X 24 X 4'  = RC036244
  36 X 36 X 6'  = RC036366
  48 X 24 X 4'  = RC048244
  48 X 36 X 6'  = RC048366

REST ROOM
  TOILET        = RRTOILET
  NORMAL STALL  = RRSTALLN
  WIDE STALL    = RRSTALLW
  URINAL        = RRURINAL
  SINK, SINGLE  = RRSINKSG
  SINK, DOUBLE  = RRSINKDB

SHELF CART
  3' X 18 X 5'  = SC036185
  3' X 24 X 5'  = SC036245
  4' X 18 X 5'  = SC048185
  4' X 24 X 5'  = SC048245
  5' X 18 X 5'  = SC060185
  5' X 24 X 5'  = SC060245
  6' X 18 X 5'  = SC072185
  6' X 24 X 5'  = SC072245
```

```
SHELF RACK
  3' X 12 X 4'  = SR036124
  3' X 12 X 7'  = SR036127
  3' X 18 X 4'  = SR036184
  3' X 18 X 7'  = SR036187
  3' X 24 X 4'  = SR036244
  3' X 24 X 7'  = SR036247
  4' X 12 X 4'  = SR048124
  4' X 12 X 7'  = SR048127
  4' X 18 X 4'  = SR048184
  4' X 18 X 7'  = SR048187
  4' X 24 X 4'  = SR048244
  4' X 24 X 7'  = SR048247
  4' X 36 X 7'  = SR048367
  4' X 48 X 7'  = SR048487
  5' X 18 X 7'  = SR060187
  5' X 24 X 7'  = SR060247
  5' X 36 X 7'  = SR060367
  5' X 48 X 7'  = SR060487
  6' X 18 X 7'  = SR072187
  6' X 24 X 7'  = SR072247
  6' X 36 X 7'  = SR072367
  6' X 48 X 7'  = SR072487
  8' X 18 X 7'  = SR096187
  8' X 24 X 7'  = SR096247
  8' X 36 X 7'  = SR096367
  8' X 48 X 7'  = SR096487

TABLE, ROUND
  3' DIA. X 30  = TR03636-
  4' DIA. X 30  = TR04848-
  5' DIA. X 30  = TR06060-
  6' DIA. X 30  = TR07272-

TABLE, STANDARD
  3' X 24 X 30  = TS03624-
  3' X 30 X 30  = TS03630-
  4' X 30 X 30  = TS04830-
  4' X 36 X 30  = TS04836-
  5' X 30 X 30  = TS06030-
  6' X 30 X 30  = TS07230-
  6' X 36 X 30  = TS07236-
  8' X 30 X 30  = TS09630-
  8' X 36 X 30  = TS09636-

TRASH CAN
  18 DIA. X 24  = TC018182
  24 DIA. X 24  = TC024242
  30 DIA. X 30  = TC03030-
```

Figure 7.1 List of blocks (*Continued*).

3. 36 = the depth of the bench in inches.

4. 6 = the height of the unit to the top of the hood in feet.

The width of the equipment is always measured across the front as the user faces it in the operating position. This means that the width of the equipment may be less than the depth, such as a file cabinet where the clerk stands in front, facing the drawers. The filename for the second file cabinet on the list is FC015244, which has a width of 15 in (facing the operator) and a depth of 24 in. The height of the file cabinet is 4 ft.

The height of the object is measured to the nearest foot, unless the height is the standard 30 in of most sit-down equipment such as benches and desks, or is less than 1 ft. If the height is 30 in it is represented by a dash, as in the BENCH, PLAIN filename BP04830-, which is 48 in wide, 30 in deep, and 30 in high.

The 3-D view of the layout does not require accurate heights as it is used only to convey a general impression of what the finished area will look like. Whether a piece of equipment is 3½ or 4 ft high will make little difference in the appearance of the 3-D view of the layout.

You will notice the terms *low* and *high* are used in this book to differentiate between the 30-in and 3-ft heights of various sit-down equipment. Chairs are denoted as low or high depending upon the height of bench they will be used with. For instance, all 30-in-high benches come equipped with a *low* chair, and all 3-ft-high benches come equipped with a *high* chair.

For general information, a 3-ft-high bench can be used in the stand-up position or with a high chair in the sit-down position. If you want to show a high bench without a chair, just explode it and erase the chair. All workbenches are shown in both the 30-in- and 3-ft-high configurations, as both are common sizes in the manufacturing industry.

The 3-ft-high benches have an extra line at each end to differentiate them from the normal 30-in-high benches. This can be used as an identifier when you are laying out an area that has a mixture of the two heights of benches and you want to group the common sizes together.

Also, you will notice that if an object is less than 1 ft high, the height is represented by a 0 in the filename. For example, CONVEYOR, FLOOR has a filename of CF240360. It is 20 ft wide, 36 in deep, and 6 in high off the floor, although the height is zero in the last character of the name. It is a very low conveyor used for transporting bulk cartons so they can be easily moved by hand.

The common manufacturing industry sizes are listed under each category of equipment. For instance, a hooded bench normally is supplied either 6 or 8 ft long and 3 ft deep. These common sizes for all the equipment blocks were derived from the equipment actually being used in the manufacturing industry.

Some examples of the 3-D equipment blocks are illustrated in Fig. 7.2. Notice the simplicity of construction of each block. They are designed to identify each category of equipment yet be simple enough to let your computer operate at maximum speed during the redraw and regeneration processes. When complicated 3-D images are used for a layout on a modest computer, they can slow it down dramatically. This is especially true when you use the HIDE command.

Since the blocks are simple, it may be difficult for the casual observer to differentiate between similar pieces of equipment. This problem was solved by assigning numbers to each piece of equipment, which will be shown on the layout.

The 3-D image can be a very important aid to people who are not familiar with plan views and cannot visualize a 2-D floor plan. Once they see a 3-D solid model drawing of the layout, they can understand the arrangement immediately. These are the major reasons for designing a plant layout in three dimensions and plotting various views:

BENCH, HOODED

BENCH, PLAIN

BENCH, TECHNICIAN

BOOKCASE

CABINET

CHAIR, LOW

CHAIR, EASY

COAT HANGERS

CONVEYOR, CORNER

CONVEYOR, FLOOR

CONVEYOR, GRAVITY

CONVEYOR, TRANSFER

DESK, FOREMAN

DESK, STANDARD

FILE CABINET

FILE, LATERAL

FORKLIFT

LOCKER

LATHE

MILL

TREE

PALLET RACK

PALLET STACK

PARTITION, OFFICE

MAN, STANDING

RACK, CABLE

SHELF CART

SHELF RACK

TABLE, ROUND

TABLE, STANDARD

Figure 7.2   Examples of 3-D blocks.

1. To provide the users with a more realistic picture of the floor plan so they can offer suggestions readily.

2. To ensure that the appearance of the work center is acceptable and does not contain any undesirable obstacles to the flow of material or the interaction of the workers.

3. To demonstrate the new floor plan to management in a form that can be quickly visualized and approved without the need for lengthy explanations.

Other Blocks

The categories of blocks that are not equipment include the door, column, fire extinguisher, people, rest room, and symbols. Most of these blocks are shown in Fig. 7.3.

DOORS

The door blocks shown at the top of Fig. 7.3 are labeled right-hand if the hinges are on the right-hand side as the door opens toward you, and left-hand if they are opposite. An example of a door on a layout is shown in Figs. 1.1 and 1.2, the layout demonstrations. This door design is the result of numerous trials to find an unobtrusive symbol that would not distract the viewer from the purpose of the layout.

The architectural door is not acceptable, as it is usually shown wide open with an arc trace to indicate the reserved area. When a group of these doors appears in a layout, it is somewhat difficult to concentrate on the flow of material because the doors are so prominent. In the worst case, it appears that the purpose of the layout is only to show where the doors are. The door blocks shown in Fig. 7.3 solve all these problems and will not intrude on the viewer's concept of the layout.

To insert a door in a layout drawing, simply follow these steps (examples are in parentheses):

1. Leave the SNAP distance at 3 in and turn ORTHO on.

2. Decide on the door width (36 in).

3. Make an opening in the wall for the door (36 in wide).

4. Find the door configuration on Fig. 7.3 (RH, 270 degrees).

5. Insert the proper door block (DR36). The base point is the hinge of the door.

6. Rotate the door block (270 degrees) until it is oriented correctly.

COLUMN AND FIRE EXTINGUISHER

The column blocks are simple extruded squares or hexagons, 9 ft high and shaded at the top so they will be highlighted in the

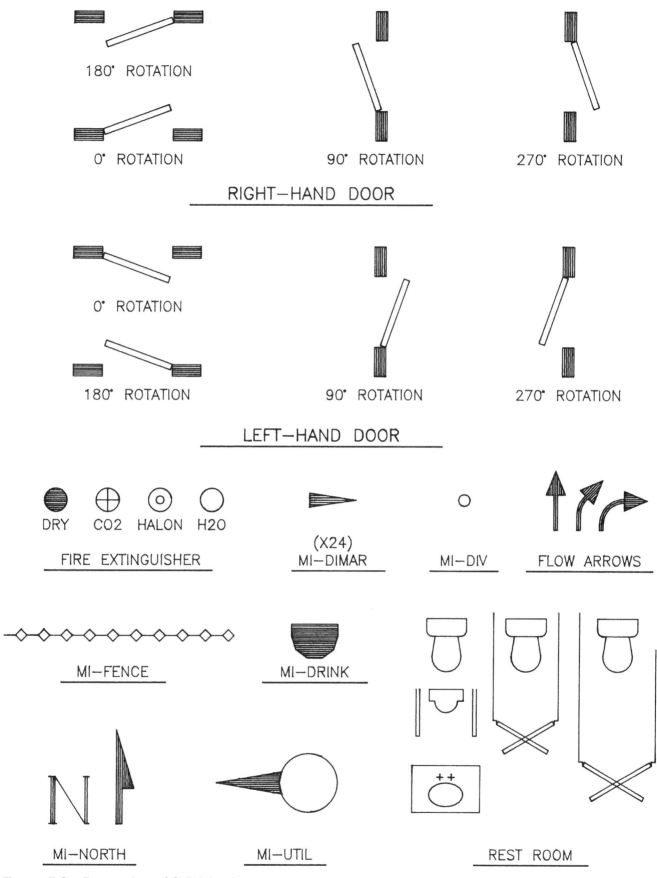

180° ROTATION

0° ROTATION

90° ROTATION

270° ROTATION

RIGHT—HAND DOOR

0° ROTATION

180° ROTATION

90° ROTATION

270° ROTATION

LEFT—HAND DOOR

DRY   CO2   HALON   H2O

FIRE EXTINGUISHER

(X24)
MI—DIMAR

MI—DIV

FLOW ARROWS

MI—FENCE

MI—DRINK

MI—NORTH

MI—UTIL

REST ROOM

Figure 7.3   Examples of 2-D blocks.

layout. The fire extinguisher blocks will appear as shown in Fig. 7.3 for the type indicated: dry chemical, carbon dioxide, halon, and water.

PERSON

The blocks for people are purposely simple and stylized. Notice the blocks for a man, standing and a man, sitting in Fig. 1.2, the 3-D layout demonstration, and how only the general impression of a man is transmitted. The blocks for man and woman, standing can be placed in key places on your 3-D views if you wish. The blocks for man and woman, sitting can be placed on the chairs next to the benches to provide a more realistic example of the finished work center appearance. The people should be placed on a separate layer so they can be turned off for the plan view layout.

These blocks of people are used only to portray the relative size of the equipment in the layout. There are third-party software suppliers for AutoCAD that provide detailed images of people that can be displayed in any position, but again the computer speed may be adversely affected. Also, your time is generally not worth the extra effort required to incorporate these images in your layout, unless you have some formal presentation requirements.

REST ROOM

The rest room blocks consist of a toilet, normal stall, oversize stall, urinal, single sink, and double sink. The two stall blocks are designed to be exploded and connected side by side. When you place them together, using a 3-in SNAP distance, the single wall of two adjacent stalls will form both sides of the wall between them. Both right- and left-hand doors are included on the blocks, so one of them should be erased. As you can see, some cleanup will be required depending upon your arrangement.

The sinks are supplied as an example since there is no standard size to this equipment. The size of the single and double sink is more than adequate for most installations.

DIMENSION ARROWHEAD

The MI-DIMAR dimension arrow block in the miscellaneous category will need some further explanation. FILL will be turned off for the layout drawing because all the blocks are composed of traces. If FILL were on, the traces would appear as solid colors and the details would be hidden. Since FILL is off, the standard dimension arrowhead will not be filled and will not delineate the object lines as well as a filled arrowhead.

The solution to this problem is to replace the standard arrowhead with a new one, MI-DIMAR, that is filled. This is accomplished by using the DIM DIMBLK command. The series of commands is:

INSERT MI-DIMAR (do not change the size or rotation)

ERASE the last entity

DIM DIMBLK MI-DIMAR

Now whenever you ask for a leader or a dimension, the arrowhead used will be the filled MI-DIMAR block. This block will also respond to the DIMASZ and DIMSCALE commands. The blank layout drawing on the diskette, LAYOUT.DWG, already has this new arrowhead installed, so no other adjustment is necessary. The routine mentioned earlier should be used if you want to install the arrowhead on a different drawing.

DIVIDE BLOCK

The MI-DIV block is used as a divider mark with the DIVIDE command. This block is sized for a drawing with a scale of $\frac{1}{8}$ in = 1 ft or $\frac{1}{4}$ in = 1 ft. To use this block, the series of commands is:

INSERT MI-DIV (do not change the size or rotation)

ERASE LAST

DIVIDE (select object) BLOCK MI-DIV NO

You will only have to insert this block once in each drawing. However, you will still have to name it every time that you divide an object. The blank drawing, LAYOUT.DWG, already has this divide block installed. The complete routine mentioned earlier should be used if you want to install the divide block on a different drawing.

CHAIN-LINK FENCE, DRINKING FOUNTAIN, NORTH ARROW

The MI-FENCE block represents a section of chain-link fence 10 ft long. The sections can be connected, they can be exploded and shortened, or a small section can be removed and used as a door. The MI-DRINK block represents a drinking fountain and is slightly oversized so it can be easily seen on the layout. The MI-NORTH block is used to indicate north and can be scaled to suit your drawing.

UTILITY INDICATOR

The MI-UTIL block is used to indicate exactly where a particular utility should be installed. It is very important for you to supply the general contractor with a drawing that is accurate regarding the utility description and location. Most of the time you will not be present when every utility is installed, and the contractor will use your drawing to direct the project. If the drawing is not clear, the workers will install the utilities anyway, as best they can.

The base of the utility block is at the end of the point, and the block can be rotated from there after insertion. A utility identifying number will be located in the circle, and the point of the arrow can be located by dimensions to the building structure. This procedure will be explained in more detail later in the book.

Custom Block Construction

There will, inevitably, be some equipment in your plant that will not be in the library of blocks supplied on the diskette. You can make your own custom blocks of virtually any equipment with a minimum of effort. It is a simple task once you have measured and recorded all the pertinent dimensions.

Figure 7.4, a blank layout drawing, is a drawing of the display that you will see on your monitor when you retrieve the LAY-OUT.DWG file. It will be explained in more detail in the next chapter. It is VIEW O. VIEW A is an enlargement of VIEW O. VIEW B is in the lower left-hand corner of VIEW O and is set aside for constructing blocks in two dimensions.

VIEW C is a 3-D image of VIEW B and is shown in Fig. 7.5, the block construction area. It is provided so you can construct your own blocks and view them in 3-D easily. The elevation posts in the background can be used to check the height of your blocks. VIEW F is a side view of the area that can be used to check the exact height of your blocks if that is important. If you inadvertently view your drawing in 3-D and cannot go back to a plan view, just enter the PLAN command and press Return for the default selection.

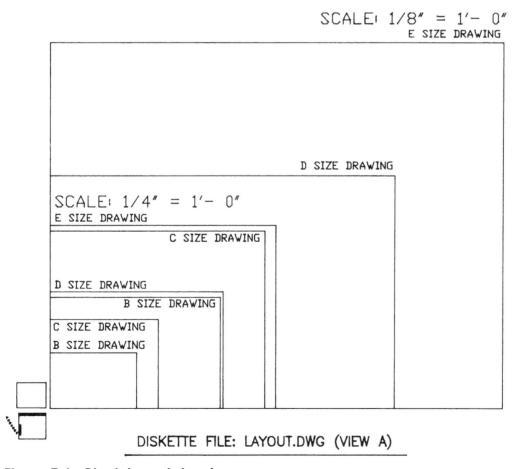

Figure 7.4 Blank layout drawing.

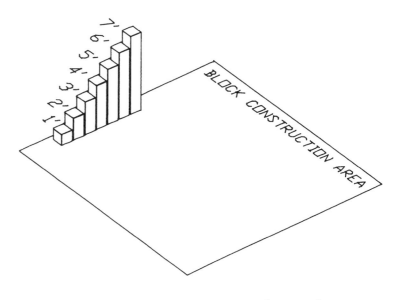

DISKETTE FILE: LAYOUT.DWG (VIEW C)

Figure 7.5   Block construction area.

The 3-D viewpoint shown throughout this book is the result of a trial-and-error process that was used to determine the optimum angle for displaying a plant layout. The final result was a VPOINT value of 10,-8,9 that you can use in answer to the VPOINT command. These are the coordinates of your eye as you look at the origin of the coordinate system. If you want to adjust your viewpoint slightly, you can vary these x,y,z values. The series of commands required to display a portion of your layout in 3-D is as follows:

VPOINT 10,-8,9 To view the entire layout, right side.
VPOINT -10,-8,9 To view the entire layout, left side.
ZOOM WINDOW To display a wire-frame view of the desired area of the layout.
VIEW SAVE Name the view for future retrieval.
HIDE To display a 3-D model view.

Use the HIDE command cautiously if you have a modest computer and have an extensive layout shown on your screen. It can take a half hour or more to hide the lines under these conditions. A high-performance computer, however, will perform the command in a very short time. The best solution to constructing blocks after you have started your layout drawing is to do the construction on a separate copy of LAYOUT.DWG. Then you can hide the lines quickly, and you will be able to see a 3-D solid model of your new blocks. After the blocks are complete, WBLOCK them to a separate file and insert them into your layout.

The 3-D blocks are primarily composed of traces. A trace can be any width, length, elevation, and thickness and can be shortened in length, but not extended. The advantage of using traces is that all surfaces are opaque after the HIDE command is used, and they are as easy to use as toy blocks.

As you record the dimensions of an object, decide how to represent it as one trace or a series of traces. Decide what the elevation and thickness of each trace will be, and record them. Remember, the *elevation* is the distance from the floor to the bottom of the trace, and the *thickness* is the distance from the bottom to the top of the trace. As an example, a 12-in cube that is on a 3-ft-high table has an elevation of 36 in and a thickness of 12 in.

All the blocks in the library have been constructed to the nearest 3 in of the actual dimension of the equipment. Rarely will you need better accuracy than that. Construct all your custom blocks to the nearest 3-in interval also. The SNAP distance on the blank layout drawing is set to 3 in, which is very quick and convenient to use during the design phase. Set the GRID to 4X for a 1-ft pattern.

All the blocks in the library will assume the color of the layer they are on, but you can change them to any color independently of that layer. That is accomplished by constructing the blocks on LAYER 0 and assigning the color BYBLOCK to the entities before they are made into a block. The entities will then respond to layer and color changes just like primitive entities (lines, arcs, text, traces, etc.).

As an example, suppose you have loaded LAYOUT.DWG and you want to make a block for an odd-sized workbench with a unit on it that is being assembled for production. Starting with VIEW B, you would perform the following steps and commands (examples and explanations are in parentheses):

1. Measure the workbench (54 in wide × 30 in deep × 36 in high).
2. Measure the unit (18 in wide × 12 in deep × 12 in high).
3. Change the computer setting to LAYER 0, SNAP 3.
4. Construct a trace for the workbench:
    a. TRACE WIDTH 30 from point 0 to point 54 (the bench).
    b. CHANGE LAST PROPERTIES THICKNESS 36 (the elevation is the default 0).
5. Draw a line at both ends of bench to indicate a height of 3 ft.
    a. CHANGE LAST PROPERTIES HEIGHT 36.
    b. CHANGE LAST PROPERTIES THICKNESS 0.
6. Construct a trace for the unit on top of the bench:
    a. TRACE WIDTH 12 from point 0 to point 12 (the unit).
    b. CHANGE LAST PROPERTIES THICKNESS 12.
    c. CHANGE LAST PROPERTIES ELEVATION 36.
7. MOVE the unit to the proper position on the bench.
8. INSERT CW018182 (high chair) into the proper position.
    a. The chair block has the color BYBLOCK built in.
9. VIEW C (to see the set of bench, unit, and chair in 3-D, wire frame).
10. HIDE (to see a 3-D solid model of the set).
11. VIEW B (to see the set in 2-D).
12. CHANGE COLOR of all entities to BYBLOCK.
    a. All parts will revert to color 7 (white).

13. Make a block out of the assembly.
    a. The filename for this custom bench is BP054303.
    b. The base point is the upper left-hand corner.

To save time, you can construct many blocks of common equipment at one time and use one 3-D routine to view them all as a group. Design the blocks so they convey the impression of the equipment, yet keep them simple. That is the same approach that an artist uses. In designing a layout, your computer is an artistic as well as an engineering tool of the trade, and your layout will reflect your artistic talent for constructing blocks as well as your engineering expertise. Notice the mill, the lathe, and the forklift in Fig. 7.2, the examples of 3-D blocks, and how only a few traces convey the impression of the equipment.

You can construct the traces in 2-D using VIEW B and then edit the elevation and thickness in 3-D using VIEW C. As you assign new height values to the trace, you will see it jump up and down and change thickness instantly. Then you can check the results by comparing the dimensions to the elevation posts in the background.

For some reason, Autodesk did not include the CHANGE PROPERTIES ELEVATION and CHANGE PROPERTIES THICKNESS commands in AutoCAD Release 11, but added it back in Release 12. If you are using Release 11, you will have to set the elevation and thickness of the drawing with the ELEVATION and THICKNESS commands, then construct each trace, and finally return them both to zero before continuing with the drawing.

If you have to put two traces side by side against each other, be sure to check the image integrity with the HIDE command. For some reason, AutoCAD will often hide the line of contact between the two traces. The line will appear to be present in a wire-frame view, but after the HIDE command is given, it disappears. To prevent this from happening, separate the adjacent traces by a 1-in space. As an example, notice the elevation posts in Fig. 7.5. They appear to rest against each other, but are actually separated by a 1-in space to ensure against line loss.

Do not draw lines on the top of a trace that terminate at its corners. Sometimes, during the HIDE command, AutoCAD will hide these lines, and they will disappear from the drawing. You can draw a cross, for instance, on top of a trace, but do not extend the lines to the corners. Set the height of these lines to the top of the trace, and set the thickness to zero.

When you measure very complicated equipment, make a sketch of the assembly of traces that will be required to make the block. Then assign dimensions to each trace. If it is a giant, like some computer-controlled machines, take an instant photograph of it for reference later, when you're sitting at your computer.

If the equipment has a noticeable overhang, then indicate that overhang with a dotted line in the block at the proper elevation in two dimensions. It is important to show the actual footprint of the equipment on the floor with solid lines. But it is also important to know about the overhang when you position the equipment on the layout. Notice the bed of the mill in Fig. 7.2. It is represented by a dashed line, indicating an overhang, but is not part of the footprint.

When the equipment is actually placed on the plant floor on moving day, the movers will use the footprint, not the overhang, to position the equipment. If the placement is critical, you can specify the exact location of the footprint on your layout. This subject will be covered again later.

If you want to use part of a block and have it fixed to a layer color, explode the block and change the color of the part from BYBLOCK to BYLAYER. If you want to use an entity of a block to make another block, explode the block but do not bother to change the entity color. The color of the new block entities will be BYBLOCK anyway.

List of Blocks

As you make blocks for your equipment, remember to record their file numbers on the list-of-blocks file (BLOCKS.WP) on the diskette. You can load this file into WordPerfect or any compatible program. If you follow the same numbering system that was used on the original blocks, the numbers will fit right into the same format and will be easy to find.

For instance, suppose your new equipment is a punch press that is composed of four traces. The designation MT is used on the list to denote machine tools, but you can use PP for punch press instead if you have many of them. If the base of the equipment is 6 ft wide, 4 ft deep, and 8 ft high, the filename will be PP072488. Make a note of the other trace dimensions to use when you actually design the block for the machine.

You can open a space for this new category after PRINTER so that it will be in alphabetical order; then enter this new block after the heading "PUNCH PRESS" in bold type. This list of blocks will keep growing with your plant layout projects, and each block will be easy to find if the list is well organized. It will take some experimentation to become familiar with how to adjust this columnar list for new entries, but it is worth the effort. To move the cursor between columns, use the command CTRL-HOME and then the right or left arrow.

When you want to insert an equipment block, just use the INSERT command followed by the filename and insertion point. Each piece of equipment will have the 3-D and color characteristics built in, ready to be displayed by your VPOINT command.

8

New Plant Drawing

The blank drawing that is used for the Flexible Division plant layout will be described in this chapter along with the directions for using it. The method of inserting all the blocks into your drawing will also be described. At the end of the chapter, the structural drawing of the new plant with the initial plan for material flow will be shown and described.

Blank Plant Layout Drawing

First, copy the blank layout drawing, LAYOUT.DWG, from the diskette to your hard disk. When you retrieve this file, the screen will look like the blank layout drawing in Fig. 7.4. This first display is named VIEW O and is purposely made small to use only half the screen. On modest computers using AutoCAD, after the regeneration is made on a reduced image such as this, later displays will be reproduced by redrawing instead of regenerating, which will speed up the computer.

As you design your layout, however, the computer may revert to regenerating every display for various reasons. To halt this slower procedure, just restore VIEW O. The display will then be regenerated, and the computer will start redrawing again. Of course, this routine is not required on later releases of AutoCAD.

To familiarize yourself with this drawing, you can restore several of the views that have been mentioned in previous chapters. These views have been assigned letters to make them more convenient to restore:

VIEW O	The initial view, Fig. 7.4
VIEW A	Closer view of VIEW O
VIEW B	2-D block construction area
VIEW C	3-D block construction area, Fig. 7.5
VIEW D	2-D layout demonstration, Fig. 1.1
VIEW E	3-D layout demonstration, Fig. 1.2
VIEW F	Side view of the block construction area

If you are unfamiliar with working in three dimensions, you can start using the layout demonstration as a learning tool. With VIEW E on the screen, move some blocks around, explode others, and change the height and thickness of the individual traces. Alternate between VIEW C and VIEW D, and notice the effect of some of your commands on the layout. Then erase the demonstration.

You will notice on VIEW O that the initial display of the file consists of a series of rectangles, called *outlines*, with text labels near them. These outlines represent the plot sizes available for various engineering paper sizes that you may use to plot your layout. The outline plot sizes used on the drawing for various paper sizes are as follows:

	Engineering paper size, in	Outline plot size, in
A-size drawing	8.5 × 11	7.3 × 10
B-size drawing	11 × 17	10 × 15.8
C-size drawing	17 × 22	16 × 20
D-size drawing	22 × 34	21 × 32
E-size drawing	34 × 44	33 × 42

There is some variation in engineering paper sizes and the plot size that your plotter can produce when using a given paper size. You can adjust these outlines to suit your system as required. The first step, however, is to determine what scale you want to use and what size paper to plot it on.

A scale of $\frac{1}{8}$ in = 1 ft is ideal for making plots of your layout during the initial phase of development. The drawing will be reasonably small and can be examined more conveniently by users than the larger sizes. Later you can use a plot scale of $\frac{1}{4}$ in = 1 ft for presentations to management and for plotting the layout as a guide to be used on moving day. On moving day, especially, you will want to plot a large-size layout so people can read it quickly as they pass by, carrying equipment.

To decide what size paper you will need, measure the overall plant dimensions and insert a rectangle on the drawing that represents the outermost structure of the plant. Place the lower left-hand corner of the rectangle at the 0,0 coordinate position. It will then be apparent which of the outlines on the drawing will be required to plot your plant at both scales. The outlines for each scale are on different layers with different colors. You can easily observe which paper size will suit your needs.

For instance, if your overall building size is 130 × 110 ft, the paper size required will be C for ⅛-in scale and E for ¼-in scale. You have to use the smaller of the two outlines for drawing the layout to be sure that everything will fit at both scales. That size will also allow you to add notes plus a legend beside your drawing. Erase all the unused outlines and text after you have chosen an outline.

If the largest paper size that your plotter can use is still too small, then decide how to divide up your plant layout so it can be plotted in sections. Also, the ⅛-in scale may require D-size paper, and you will have to divide the plant so you can plot it at ¼-in scale. This process of assigning paper sizes may seem premature right now, but it can save a great deal of time and inconvenience later when you are locked in to some configuration and find that you cannot plot it easily.

Settings for Blank Drawing

The settings and other notes for the blank layout drawing are shown in Fig. 8.1. Most of the settings are conventional, but explanations are shown for some items:

3. LAYER NAMES.   The layer names indicate how the layers are used in the layout demonstration. Some of the layer names will also be referred to later in the book. The FILL and NUMBERS layers are mentioned in macros in later chapters. Other layers can be used for the various work centers of your plant. If you assign colors that can be plotted to these layers, the preliminary layout plots can be color-coded to delineate the work centers more clearly.

4. VIEWS.   It is much easier to move around a large layout if you divide it into views and keep a list for easy reference. By assigning letters to each view, you can restore them easily. This is especially important for 3-D views, which are difficult to display accurately. After you adjust the display to show exactly the right 3-D view of a section of your layout, be sure to assign a name to it.

6. TEXT STYLE.   Because this drawing is also filed in DXF format, all text styles are STANDARD with a TXT font. If you are going to convert your drawing back to a DXF format later, do not use any other style of text.

11. OTHER PARAMETERS.   The circle percent is set to 1000 so circles will appear round and smooth. Hexagons will be used for circular columns and other round 3-D constructions in the layout. It will be difficult to differentiate between circles and hexagons if the circle percent is much less than 1000.

12. SCALES.   You should use the ¼ in = 1 ft settings only if you never intend to plot your layout at ⅛ in = 1 ft. If you use the ¼-in settings and then plot at ⅛ in later, the text will be too small to read easily. Normally, the ⅛ in = 1 ft scale settings will be used for the entire plant layout including the text. Then later,

The settings and other guides that pertain to the blank plant layout drawing file on the diskette are shown below:

1. FILENAME.   LAYOUT.DWG

2. SCALE FOR PLOTTING.   ⅛ in = 1 ft or ¼ in = 1 ft.

3. LAYER NAMES.   If you want to change the names, note that layers FILL and NUMBERS are included in the macros described later in the book. The current layer is set to WALLS. The layer names used in the demonstration are as follows:

    0   (gra 8) = To create blocks.

    WALLS   (wht 7) = Walls, doors, etc.

    FILL   (blu 5) = Hatching for the walls (used in macro).

    EQUIP   (red 1) = All equipment.

    PEOPLE   (grn 3) = Figures of people and units on conveyor.

    NUMBERS (yel 2) = Equipment numbers (referenced in macro).

    1-8   (cyn 4) = Outlines to plot the scale ⅛ in = 1 ft.

    1-4   (vio 6) = Outlines to plot the scale ¼ in = 1 ft.

4. VIEWS.   These views can be restored with the VIEW command:

    O   View when the file is initially retrieved.

    A   View that is closer than view O.

    B   2-D view of the block construction area.

    C   3-D view of the block construction area.

    D   2-D view of the layout demonstration.

    E   3-D view of the layout demonstration.

    F   Side view of block construction area.

5. BLOCKS.   The library of blocks is not included in the file, but the block named LIBRARY.DWG can be easily inserted.

6. TEXT STYLE.   STANDARD text style with the TXT font is assigned to all text.

7. LINE TYPE.   All lines are continuous.

8. MENU.   ACAD.MNU is loaded when the file is retrieved.

9. UNITS.   Architectural, 1'-3", denominator = 1, east 0 angle direction, CCW, decimal degrees, decimal places = 0, units = 1 in.

10. LIMITS.   Set to include the largest drawing outline.

11. OTHER PARAMETERS.   Miscellaneous assigned settings:

LTSCALE	= 80	REGENAUTO	= on
ORTHO	= on	QTEXT	= off
SNAP, ON	= 3 in	FILL	= off
BLIPMODE	= off	DRAGMODE	= auto
FILLET RADIUS	= 0	OFFSET	= 6 in
FAST ZOOMS	= yes	CIRCLE PERCENT	= 1000
UCSICON	= off	UNDO	= all

Figure 8.1   Settings for layout drawing.

12. SCALES. Change these settings according to the scale that you will be using for each drawing. They are initially set for ⅛ in = 1 ft scale.

Drawing scale	⅛ in = 1 ft	¼ in = 1 ft
TEXT height	10 in	5 in
DIMSCALE	1	1
DIMASZ	12 in	6 in
DIMTXT	10 in	5 in
DIMEXE	6 in	3 in
DIMEXO	6 in	3 in

13. PAPER AND PLOT SIZES. These dimensions correspond to the paper size and plot size represented by the outlines shown on the screen. The equivalent dimensions, in feet, are shown for the outlines at both scales.

	B-size paper	C-size paper
Paper size	17 × 11 in	22 × 17 in
Plot size	15.8 × 10 in	20 × 16 in
Outline, ⅛-in scale	126 × 80 ft	160 × 128 ft
Outline, ¼-in scale	63 × 40 ft	80 × 64 ft

	D-size paper	E-size paper
Paper size	34 × 22 in	44 × 34 in
Plot size	32 × 21 in	42 × 33 in
Outline, ⅛-in scale	256 × 168 ft	336 × 264 ft
Outline, ¼-in scale	128 × 84 ft	168 × 132 ft

For example, if your plant is 130 ft wide and 110 ft deep, then you can plot your plant layout on C-size paper at ⅛-in scale or E-size paper at ¼-in scale.

Figure 8.1 Settings for layout drawing (*Continued*).

if the larger-scale plot is needed, the text will be plotted at the larger scale also. The LAYOUT.DWG file has the ⅛ in = 1 ft settings preloaded.

13. PAPER AND PLOT SIZES. The outline figures shown previously are identical to those used to construct the outlines shown in VIEW O. The plot size indicates the actual area of the paper that is available to plot the layout. After you arrive at the proper outline, as described at the beginning of this chapter, you can retain it on your drawing as a guide to ensure that all of your layout within those boundaries will be plotted.

Block Insertion File

All the blocks in the library are listed in Fig. 7.1 and shown in Fig. 8.2, the block insertion file. They are included on the diskette as a single block file named LIBRARY.DWG. Whether you choose to make your drawing on your company's standard drawing template or on the LAYOUT.DWG file, you can insert all the blocks with one command.

Insert the LIBRARY.DWG block on layer 0. The base point is in the upper left-hand corner, and the block will appear as if the two sheets of Fig. 8.2 were side by side. You can use 0,0 as the insertion point, and the block will appear right below the outlines. At that point, all the blocks will be entered on the block list, and each individual block will be an entity with its own filename. As the block file loads, the duplicate blocks that are used for the demonstration will be ignored.

You can incorporate the blocks into your drawing by exploding the equipment block and then moving or copying each piece of equipment from the group into the layout drawing area. The easiest way, however, is to erase the LIBRARY block, select the block name from the list shown in Fig. 7.1, use the INSERT command, and type in the name manually. The insertion point is usually in the upper left-hand corner of the block. The block will assume the color of the layer that it is on, but you can change its color or copy it to other locations.

There are two blocks included on the LAYOUT.DWG that are used for the layout demonstration and are not needed in your layout drawing. Their names are 2 and 3. They can be deleted from your drawing after you have erased the layout demonstration. A block named 1 is used for the elevation posts in the block construction area. All these blocks were constructed so the DXF file would be more compatible with other programs.

Plant Measurements

The Flexible Division IE determined the floor area required for the new plant by using the values calculated in Fig. 5.1, the new plant equipment list. The plant of their choice is shown in Fig. 8.3, the new plant drawing. It will require some modification to satisfy their needs, but in general the floor area is adequate, and the building is next door to the parent corporation.

When you make plans for moving or tearing down the walls of the new plant, be certain that they are not structural walls. If you cannot avoid this, the building structure will have to be modified to suit the new plan. In most cases a simple inspection will reveal whether the wall is structural or not. Usually, a building that is supported by columns does not have internal structural walls. That is the case with the Flexible Division's new plant. If there is any doubt, ask the architect, maintenance supervisor, or, as a last resort, a commercial building inspector.

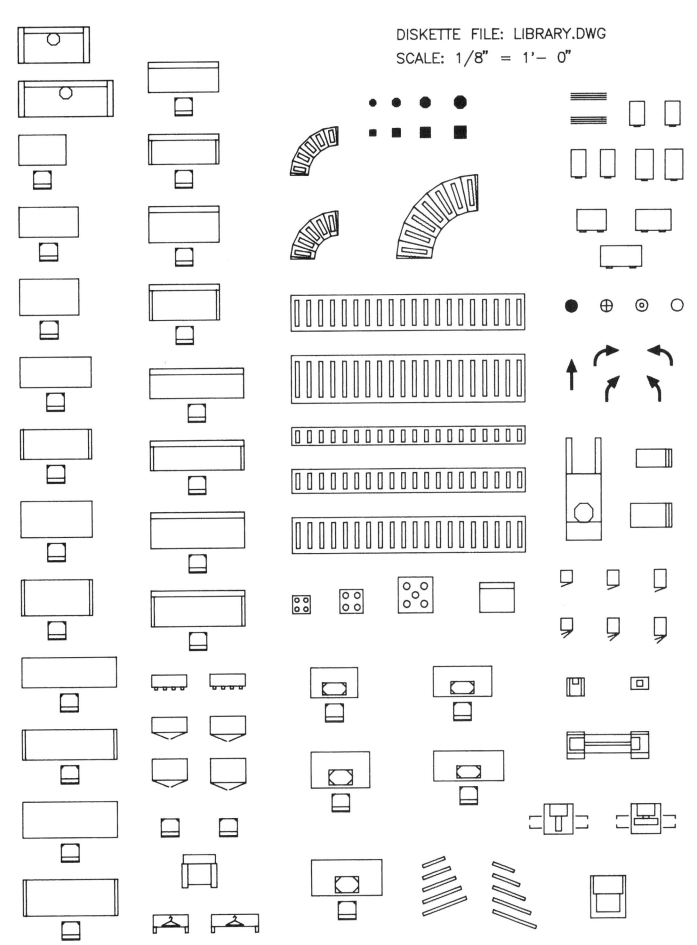

DISKETTE FILE: LIBRARY.DWG

SCALE: 1/8" = 1'- 0"

Figure 8.2   Block insertion file.

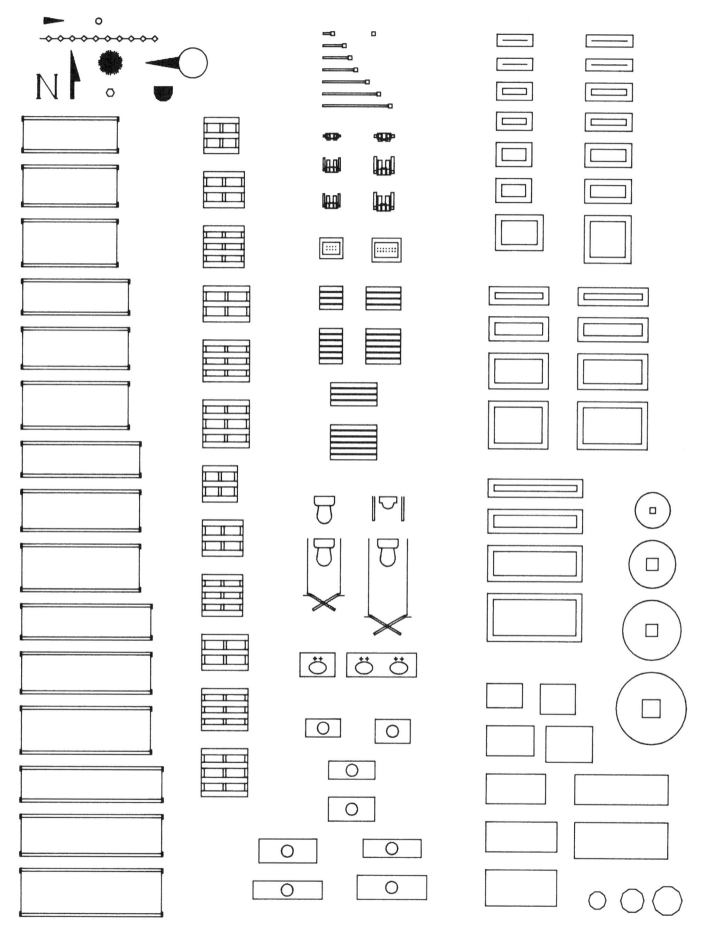

Figure 8.2   Block insertion file (*Continued*).

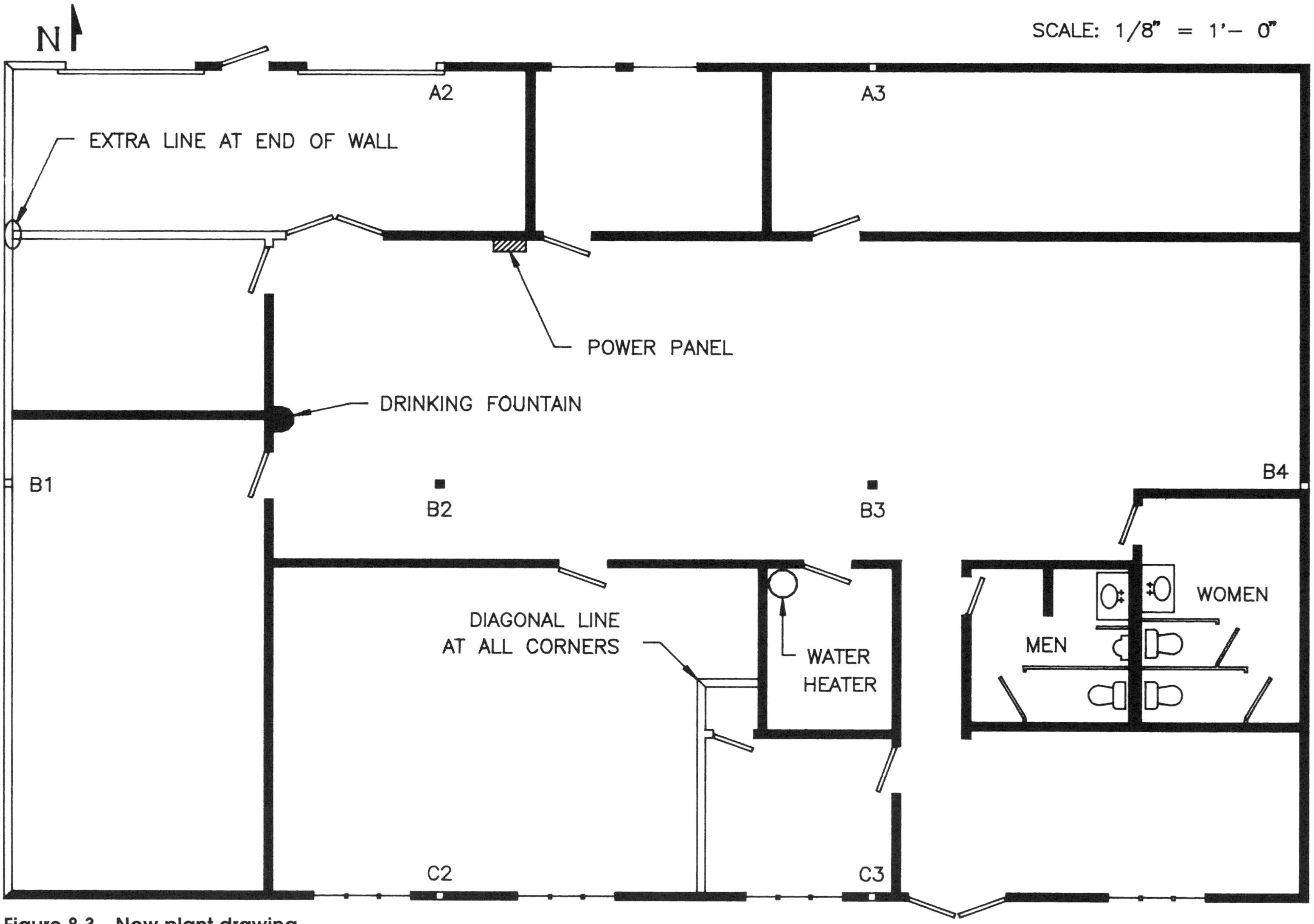

SCALE: 1/8" = 1'- 0"

EXTRA LINE AT END OF WALL

POWER PANEL

DRINKING FOUNTAIN

DIAGONAL LINE AT ALL CORNERS

WATER HEATER

WOMEN

MEN

A3

A2

B1

B2

B3

B4

C2

C3

N

Figure 8.3　New plant drawing.

The next step in the plant layout project schedule (Fig. 6.1) is task 6, *IE: Draw new plant structure.* The IE constructed the drawing of the new building and indicated the proposed material flow pattern to prepare for the first project planning meeting. During the meeting, the new plant drawing, at ¼-in scale, will be the vehicle for directing the discussion and recording the various ideas for flow improvement that will be suggested by members of the team.

Measuring the plant structure is the most arduous part of this step. As recommended before, do it yourself and avoid mistakes. An extra person to hold the end of the tape is very helpful, but *you* should do the actual measuring and tape reading. It is not difficult to measure everything yourself if you have these tape measures: one 30-ft-long, 1-in-wide tape and one 50 ft long. The 1-in-wide tape is relatively inflexible and can be guided into inaccessible locations or pushed along the floor quite easily.

If you make errors while measuring the plant structure, they can haunt the entire project right up to moving day and then torpedo it. Double checking is the best way to avoid measurement errors, and you should do so repeatedly, especially at the end of your session. Measure the major dimensions of the building again just before you leave the plant.

If you intend to design many plant layouts and want a reliable double check on your measurements, electronic distance-measuring devices called *distance meters* are available. The most accurate type that is reasonably priced uses a target about the size of a cigarette pack placed at one end of the distance to be measured. The zero reference point is at the back of the target, so it can be placed against a wall to establish the starting location. The measuring device or sender can then be positioned at any interval up to 200 ft away from the target.

When you press the signal button on the sender, the distance from the sender to the target is measured and appears on a digital display on the sender. The distance is measured with an infrared, laser, or ultrasonic beam. The reading error is about 2 in for 100 ft, which falls within the acceptable tolerance for plant layout design. All your measurements will be to the nearest 3-in interval anyway.

The measuring device can be used as an excellent verification tool for continuous measurements. If you measure many small in-line sections of the plant and can see your starting point from the end point, you can check the sum of these dimensions easily. If you want to check the column spacing of a large room, the production area, for instance, then just place the target at the first column, walk the length of the area, and take a running distance reading at each column.

No matter what procedure you use to measure the plant, verify your dimensions often. If you have an architectural drawing available, use it as a verification tool only. Do not make the mistake of assuming that it is a correct representation of the actual plant structure. Changes are made to almost every building during construction and are not recorded on the original drawing.

Consider it a challenge to locate these changes and record the dimensions properly.

To start measuring, tour the important areas and decide how they will be sketched and the dimensions recorded. Keep all the sketches in your notebook. Take instant pictures of any structures that you will need to draw in detail later, and keep the pictures in your notebook, too. You will refer to these original drawings and photographs often during the layout design process.

All your measurements will be made to the nearest 3-in interval. The walls will be either 6 or 9 in thick. Their distance from the columns will be drawn to the nearest 3-in interval. The odd-sized door widths will be set to the nearest 3-in interval. The blocks are constructed to the nearest 3-in dimension. The SNAP distance on the layout drawing will be set to 3 in. By using this universal 3-in distance, it is very simple to place any structure or object on your drawing. You will rarely have to use the OSNAP command, and equipment can be placed in exact location even though you are observing the layout at a low magnification.

You will seldom need greater accuracy than 3 in. However, it could happen if several pieces of equipment appear tight between two walls. In that event, just measure everything very closely, and make sure that all the pieces will fit.

Sketch your floor plan as neatly as possible, and allow enough room for detailed dimension notes where they will be needed. Break the building down into easily managed sections, and label the sketch of each section so you can identify it later. Save everything in your notebook. Conscientious organization of every step will save time in the long run and provide a foundation for improvement on the next layout project.

The columns are the base point reference for all other structures of the building. Therefore, the center distance of the array of columns is the first item to be measured and verified. Usually, the columns will be placed in a standard distance array on even-foot centers, but not always. Sometimes a column will be placed at an odd center distance at the edge of the array to provide for some other structure, such as an angular wall. Measure all the center distances, and disregard slight variations of less than 3 in so you can place all the columns at a uniform distance from each other on your drawing.

After the columns are placed on your sketch of each area, locate all the surrounding walls relative to the columns. Measure all door openings and their distances from a part of the structure. Locate power panels, groups of pipes, or other encroachments that will affect the area available for the work centers. Locate any equipment that will remain in place after the layout is complete. In general, imagine yourself drawing the plant structure on your computer, and then make sure that all the information you will need is noted on your sketches.

The windows are important for illumination or observation from either side. Also, they should present a favorable appearance from the outside of the building, without the unattractive display of equipment, such as cabinets, placed right in the window openings.

Measure the width of the glass and the dimensions of the framing. Make a sketch of the window as it will appear on the layout drawing. Notice in the new plant drawing of Fig. 8.3, how the windows in the front of the building are indicated by simple lines and the framing is indicated by squares. Locate the windows to the adjoining walls, to the nearest 3 in.

Notice the power panel on the internal wall to the north. The width is accurate, but the 3-in depth is shown as 6 in to be more visible. It is hatched at 45 degrees with 3-in spacing. If you have many of them, place an additional shaded area around each one to indicate the access space for the electrician. Also, measure the width and depth of the columns, and record whether they are round or square.

New Plant Drawing

At this point you have retrieved the LAYOUT.DWG file, experimented with the 3-D view, familiarized yourself with the art of drawing in three dimensions, and erased the layout demonstration. You have chosen the proper outline for plotting your drawing and erased the unneeded outlines. You have loaded the library of blocks, and they are now at your disposal. The SNAP distance is 3 in, the grid is set at 12 in and turned on, and ORTHO is on.

In the new plant drawing, the IE used simple yet effective techniques, and the result conveys the important aspects of the building without distracting the viewer with unnecessary details. All the items are on the WALLS layer except the hatching for the walls. Here is a description of the important items:

1. *Columns.* There are only two columns in this building, and they were drawn first. If you use the LAYOUT.DWG file, all the drawing coordinates will be in the upper right-hand quadrant, so they will be positive. Use a column block to insert your first column, and place it on an even-foot, grid dot location in your drawing. Then use the ARRAY command to place the others.

2. *Column labels.* Notice in Fig. 8.3 that the internal columns are labeled B2 and B3. Since there are only two exposed columns in this building, it is not absolutely necessary to label them, but the labels are shown as examples. The columns that are hidden in the wall or are only implied are shown also. Sometimes a hidden column is enclosed, and the enclosure protrudes from the wall. Be sure to include that protrusion on your layout.

If your building has many columns, they have probably been labeled, and the labels will also be on the architectural drawing. If they are not labeled in some manner, assign alphanumeric labels to them, starting with the column that may be hidden in the upper left-hand corner of the outside wall. Hidden columns may or may not be shown on the layout.

Have your maintenance department apply labels to the columns on all four sides. This will save a great deal of time and confusion as you design the layout, and even more if the building is modified. These labels will also be a big help on moving day, enabling you and everyone else to orient your drawing to the proper columns.

3. *Walls.* Notice the walls in the upper left-hand section of Fig. 8.3, which are left unshaded to demonstrate the typical wall construction technique. All the walls are constructed with lines so they can be easily hatched by using the WINDOW method. When a wall terminates at another wall, an extra line must be added as shown, to ensure that both walls will hatch properly.

You can hatch a wall only if every line forming it is exactly on its border. That is, do not use a line for the border that extends beyond the area that you want to hatch. All walls are then hatched on the FILL layer with longitudinal lines at 1-in spacing. This layer can be frozen to maintain computer speed while the actual layout is being designed.

4. *Doors.* Door blocks were used to insert the hinged doors. The method of insertion was described in the previous chapter. The roll-up doors in the upper left-hand section are rectangles, not shaded, 3 in deep with a thickness of zero.

5. *Windows.* The windows in the north wall of the building are indicated by lines only, but the front windows have two frames in each one. Normally, most of the windows in your plant will be of uniform design. You can design one window, make a block of it, and use it for all window locations.

6. *Rest rooms.* The rest room blocks are used to indicate all fixtures. Notice that an oversized stall is used if only one is required. Later, an easy chair will be added to the women's room.

7. *Miscellaneous.* The north arrow is shown and has been scaled to suit the size of the drawing. The scale of the drawing is shown in the upper right-hand corner. In this same location, add the revision date of the drawing, and update it with each revision that you make.

Before the project planning meeting is held, the proposed location of each work center should be added to the new plant drawing. In Fig. 8.4, showing the proposed material flow, the IE assigned the work center areas according to the general orientation indicated on Fig. 5.3, the relationship diagram. In the plant layout project schedule (Fig. 6.1), this step completes task 5, *IE: Assign work centers to areas*, and task 6, *IE: Draw new plant structure.*

The initial proposal for the material flow has also been added to Fig. 8.4. The IE indicated the location and direction of the flow arrows on this drawing only as a starting point to be presented for discussion at the project meeting.

Indicate the flow of material by using the flow arrow blocks on Fig. 7.1, the list of blocks. Even though it may be obvious

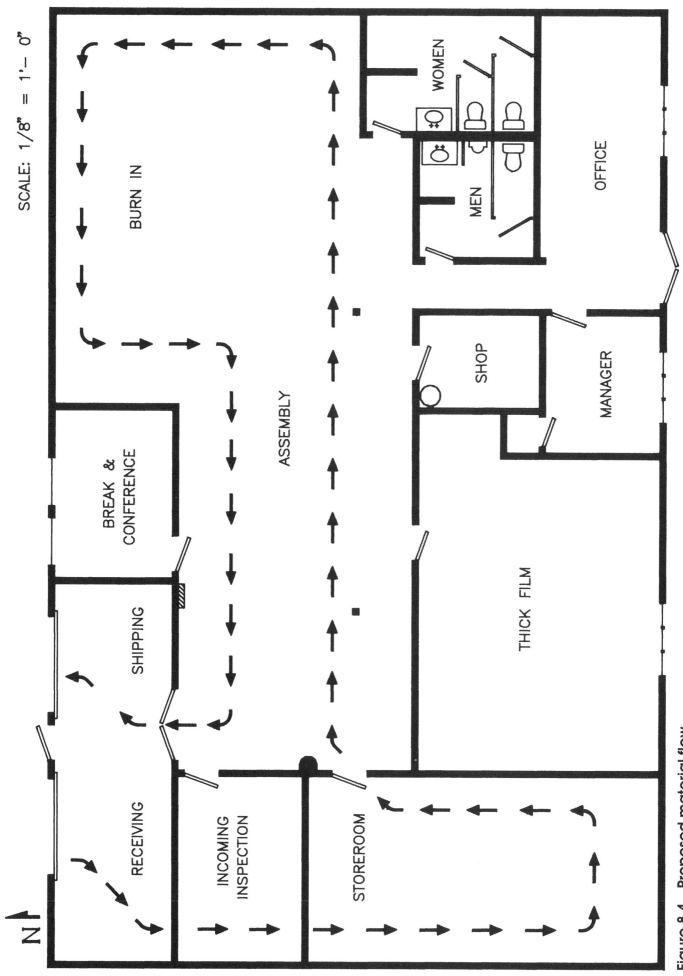

SCALE: 1/8" = 1'— 0"

N

WOMEN

MEN

OFFICE

SHOP

MANAGER

THICK FILM

BURN IN

ASSEMBLY

BREAK & CONFERENCE

SHIPPING

RECEIVING

INCOMING INSPECTION

STOREROOM

Figure 8.4 Proposed material flow..

where the work centers should be located and how the material should flow, regard it as a proposal, to be considered and accepted by the project team.

To plot your new plant drawing for the project team to review during the plant layout meeting:

1. Set the limits to the edges of the outline.

2. Plot to the limits.

3. Set the origin and size for your drawing.

4. Set the scale to: 1/8"=1'.

5. Plot the drawing.

During the meeting, leave the door open to suggestions by the project team members regarding the improvement of the system. Not only will you receive some valuable information and ideas, but also everyone will play an important role in the design of the new system and be a contributor to its successful completion.

If you view the new plant drawing from a 3-D perspective at this stage of development, all the entities should appear to be on the floor at zero elevation, except the columns which should be 9 ft tall.

9

Meetings and Macros

In this chapter, business meetings in general will be discussed and the Flexible Division layout project meeting described. Many macros for accelerating the layout drafting process will be explained, and described as they appear on the diskette.

Meeting Qualifications

Although the Flexible Division IE found it necessary to hold a layout project meeting, most business meetings cannot be justified. In truth, business meetings are the curse of the manufacturing industry. If all business meetings were banned, the productivity of all the managers, supervisors, and other participants would soar to unprecedented heights. In an eight-hour day, without any meetings, most of these people will be fortunate to perform six hours of effective work. To rob them of one or two hours from that time every day to attend a meeting of dubious value is the most detrimental burden that can be placed on any employee.

Of equal importance is the direct cost to the company, which can be hundreds of dollars in salaries and benefits for each hour of meeting. Not only that, the company has hired the meeting participants in expectation of making more money from their efforts than they are being paid. In that case, the meeting essentially deprives the company of that day's profit for those people.

The meeting schedule for most conference rooms is usually full by the start of the day, and occasionally runs beyond normal working hours. If an average of six people attend these meetings, it is the equivalent of hiring six people, full time, at an elevated salary, to sit all day and discuss elementary information that should have been handled one-on-one in someone's office.

If meetings were banned, the organizers would be forced to substitute some other form of communication that would invariably be more efficient. A very simple memo could be released on the interoffice network to the recipients, who could respond in a suitable manner. Soon, everyone would learn to communicate freely in concise messages that require no more than a few minutes to complete. Having a written record of each person's opinion is just another valuable plus.

It is the author's opinion, based on spending an entire career in the manufacturing industry, that all business meetings can be broken down into these four categories:

30 percent	Meetings of zero value to the participants. The information presented is forgotten within an hour.
30 percent	Meetings with marginal value to the participants. The information is sometimes remembered for several days. No decisions are made.
30 percent	Meetings with some value to the participants. The information is sometimes remembered for a week. If decisions are made, they have a minor impact on the future of the company.
10 percent	Qualified meetings with constructive value to the participants and the company. The information presented is recorded and used to advance the company's position in its field.

The 10 percent figure may be overly generous. This appraisal suggests that the information promulgated in 90 percent of the meetings is forgotten within a week. In general, most meetings are held as a convenience for the organizer and are a waste of time for the attendees. Memos, not meetings, should be used to disseminate information. When conference rooms are converted to offices, the result is better space utilization and increased productivity.

A meeting must have an agenda and satisfy a few basic conditions before it can be qualified:

1. The participants of the meeting must discuss the subject among themselves and come to some definite conclusions that will make an impact on the company's business. If the attendees are merely reporting to the organizer about the subject in a status meeting, then the meeting does not qualify. These people should use memos for that function instead.

2. The meeting must not be held on a regular schedule. Production priority status meetings are the worst case. Regular meetings are held to treat the symptoms of a problem. Solve the problem, and the meetings will not be necessary.

3. The organizer cannot imagine any method of addressing the subject without holding a meeting. In most cases, however, a one-on-one discussion will produce a more sincere and complete response from each person. The response will also be free from the political considerations present in every meeting.

The Flexible Division IE was fully aware of the importance of qualifying the layout project meeting and preparing an agenda for it. The agenda listed all the participants and the reports that they would present. All the participants were members of the vendor qualification team and were familiar with the problems to be discussed. The meeting was required because the participants wanted to solve the problems as a group and reach a definite plan of action for incorporating the results in the plant layout.

The IE realized that it would be more efficient to hold a meeting than to contact the participants individually. If a meeting were not held, the total person-hours involved would be greater and the plan would produce inferior results.

Usually, this is not the case. A meeting of six people can consume more than 6 person-hours, considering the extra time lost in adjusting their busy schedules. The organizer can perform the same function by talking alone with the key individuals for just a few minutes.

Plant Layout Project Meeting

Several days before the Flexible Division's layout project meeting, each attendee received the agenda. This meeting is task 7, *IE: Organize meeting to coordinate project*, in Fig. 6.1, the plant layout project schedule. Here are the leaders' introductory comments for each discussion:

1. QA manager: "The operators are making excellent progress and coming up to speed in inspecting their own work. Most of the assembly operators have been qualified, and the rest soon will be. When we move to the new plant, I have full confidence that the production personnel can be relied upon to produce a zero-defect line of products. Of course, new employees will have to be monitored very closely, but I see no problems as long as we hire the same level of responsible individuals that we have in the past."

2. Chris, the manufacturing engineer: "The fixture and equipment for testing the flash memory module have been designed, and two test stations are being built by an outside vendor. One will be used in-house, and the other will be on loan to the memory module vendor. The vendor is cooperating fully with our failure analysis program, and they will supply their own pretest burn-in oven at their outgoing inspection station. In just a few months, our incoming flash memory packs will exhibit close to zero-defect quality. Eventually we will be able to eliminate the burn-in operation entirely. I have ordered the laser trimmer, and it should arrive on schedule." This refers to task 3, *CHRIS W.: New laser trimmer*, in Fig. 6.1.

3. IE: "The new manufacturing system will be based on an in-line method of assembly designed around a gravity conveyor. All the rolling carts and bins will be the same size and color. To maintain our flexibility, we should not tailor anything to the

product. There will be one cart for each assembly bench. Each cart can hold 12 bins, and each bin will hold a kit for one assembly operation. The 107 needs six assembly operations so six different carts will be packed with 12 bins each. The bins, the carts, and the conveyor pallets will all have color-coded inserts for each model of memory bank. Usually, one product model will be assembled at a time.

"Maximum flexibility has been designed into the new system. The product being assembled can be changed often, or one product can be run continuously. The line can change to a hot product within 15 minutes. To do that, the carts for all assembly operations of the hot product will be packed in the storeroom and moved to the line. The last cold product at the first assembly bench will be completed and passed down the line to the end. All partially used carts for the cold product will be moved to the staging area for later use. The loaded cart for the first operation of the hot product will be placed at the first bench, and assembly will start right after the last cold product unit is completed.

"The wait time between assembly operations will be drastically reduced because the product will be assembled in line instead of in batches. As a result, the WIP and flow time will also be reduced. The manufacturing flow time for the 107 will be less than half of what it was using the batch method. It used to take about two weeks to complete a 107, and now it will take about four days. That time can be cut to one day if we can eliminate the burn-in cycle. Then the finished-goods inventory could be eliminated, and our capacity could be doubled just by adding another shift."

4. Pat, the storeroom supervisor: "The shipping, receiving, incoming inspection, and storeroom should all be arranged in line and connected by a gravity conveyor. QA wants incoming inspection to be walled off with locked doors. The inspection area can be somewhat smaller than it is now because a lot of our material will no longer be inspected. However, QA still wants to log all material in and out for the time being. Here is a list of the storage racks I will need for the storeroom, based on our new, reduced inventory levels. The number of supply carts for production is also on the list. I have started to organize the parts and shelves in kit picking order for the new storeroom." This report refers to task 1, *PAT K.: Plan storeroom and parts*, in Fig. 6.1.

5. Buyer: "Some minor vendors are qualified, but the memory pack supplier cannot start inspecting in-house until they receive the test equipment. The vendor who supplies all our sheet-metal parts has made a fixture for checking hole locations right at the punch and brake stations, so our assembly problems in that area are probably over."

6. Gerry, the production supervisor: "The pilot assembly line is operating as well as possible under the circumstances. They are using rolling carts now until they get the new conveyor. The operations are adequately balanced, and the burn-in test station is now operating a full nine hours per day. The goals shown in our labor analysis will probably be met after the new layout is in

place and the manufacturing system is up to speed." This completes task 4, *GERRY B.: Set up prototype assy. line*, in Fig. 6.1.

If you conduct a plant layout project meeting, encourage the participants to discuss each other's presentations. At key intervals, take time off and call upon each attendee to express an opinion of the issue under discussion. Some may have excellent contributions to make but will not offer them in a meeting where everyone is talking at once. Everyone should have an opportunity to participate in the new manufacturing system design and become a contributor, participator, and supporter.

When the meeting is over and the planning is finally complete, you can begin preparing your CAD system to draw the new plant layout. The rest of this chapter will present some helpful aids for using your CAD system more efficiently.

CAD Acceleration

Any advanced CAD program can be customized to provide a much higher rate of command execution than is available from the manufacturer. For this discussion, the AutoCAD program is used as a demonstration program, as it is very common to the manufacturing industry and is easy to customize. The instructions for customizing AutoCAD are shown in the manual, but they only describe how to enter a few basic commands. To customize the program so it performs at a truly elevated speed requires more study and an investigation of your command usage routine. It also requires a word processing program to edit the AutoCAD menu file, ACAD.MNU.

The AutoCAD screen menu was originally designed by Autodesk as a guide only, to present all the commands in a structured format. Autodesk assumed that everyone would eventually construct a customized menu. Most users, however, adopted the screen menu as a tolerable method of issuing commands. The disadvantage of this cumbersome method of constantly flipping from menu to sub-menu was overlooked by the beginner and finally accepted after it was memorized. There are only a few customers who design and use a detailed custom menu, and Autodesk refers to them as *power users*.

The use of the standard screen menu has now solidified, and almost all users are locked into this relatively slow and distracting method of issuing commands. The digitizer tablet template supplied with AutoCAD is only a modest attempt at streamlining command execution, but it can still be used as a good place to start. You can use it to test macros and become familiar with the command symbols. From the author's experience, a fully customized digitizer and screen menu can reduce the AutoCAD drawing time by 50 percent, possibly more.

If you use a fully customized menu, the amount of time that you save will depend on how much of your time is spent designing versus drawing. That is, if you spend 60 percent of the project time thinking about the design and 40 percent drawing, then the total time saved will only be 20 percent. However, if you are editing a drawing, which is almost all drawing time, then you could finish the project in half the time.

There is another advantage to using customized menus that may elevate the quality of your output and reduce your mental fatigue. If you use a customized menu, you will not constantly change your train of thought from concentrating on the design to issuing commands. The screen menu takes not only more time but more concentration as well.

Almost every command has alternatives that must be selected before it will execute properly. This selection process redirects your thoughts from the design to command functions. For instance, to change a text string requires several selections for height, style, etc., before AutoCAD asks what new text you want to write. By the time you start writing the new text, your design thoughts have been interrupted, and they must be instantly recalled before you can start designing again.

This constant thought interruption causes subconscious mental fatigue, eventually causing your output quality and rate of command issue to deteriorate. This is true of all advanced CAD programs. In contrast, if you use a customized menu for inserting text, AutoCAD will skip through the unneeded alternatives and ask you directly for the new text. Your mind thus remains on the design function full-time. The title written into each box on your tablet will eventually represent the complete programmed command, and you will often forget the basic AutoCAd commands from which it was derived.

The line of characters on the menu, called the *menu item*, is used to program each box on the tablet menu. You can program each menu item and label each box to suit your exact needs. For example, the COPY command is usually presented as a single item that requires several choices before it can execute. However, you can program it to have 20 alternative forms, each ready for instant execution:

COPY (ORTHO on)	COPY (ORTHO off)
MULTIPLE OBJECT	MULTIPLE OBJECT
OBJECT	OBJECT
2 OBJECTS	2 OBJECTS
MULTIPLE WINDOW	MULTIPLE WINDOW
WINDOW	WINDOW
3 OBJECTS	3 OBJECTS
MULTIPLE LAST	MULTIPLE LAST
LAST	LAST
4 OBJECTS	4 OBJECTS

For plant layout work, each of the above alternatives can be customized to turn SNAP on and skip alternatives during command issue. The two-, three-, and four-object alternatives are used to pick entities out of a group that cannot be windowed.

To offer this wide variety of alternatives, a tablet must have a high command density similar to Fig. 9.1, which shows a digitizer template. It is a snapshot of the author's template, which is regularly revised to provide ever-greater speed.

A *space* is defined as a single location for one macro or one line item. A full-size example, 0.40 in wide, is shown at the bottom of Fig. 9.1. A box may be composed of one or more spaces but only represents one macro. Notice these features:

1. *Menu area 1.* These commands are used constantly, so the boxes are larger and more loosely packed.

2. *Menu area 2.* These boxes are used to select layers during command issue and can be used to set any layer instantly by picking the box in the right-hand column of the menu area. Layers are named by numbers and then described on the lines in the screen pointing area.

3. *Menu area 3.* These commands are used occasionally, so the boxes are smaller and more densely packed.

4. *SNAP.* The SNAP aspect ratio used on the drawing of this tablet is 0.2 × 0.05, so the text can be placed in the center of each box easily.

5. *Screen pointing area.* This area should have the same aspect ratio as the drawing area on your monitor including the screen menu area on the right. The horizontal dimension should be at least 9 in so you can draw on the screen at a reasonably low zoom factor. You can also use this area on your template to record any important information that you use constantly, such as pen color assignments or figures of doors at the four rotation positions.

6. *Columns and rows.* Numbers have been assigned to the columns, and letters assigned to the rows outside of the active area, just as in the AutoCAD standard template. Each item on the menu file is preceded by its location, [B10] for instance.

7. *COPY routine.* The series of 20 boxes mentioned in this chapter is located in rows A, B, C, and D and columns 2 through 7. A similar MOVE routine of 14 boxes is located just to the right of the COPY routine.

8. *Command title.* If the title must occupy more than one space, use several. Many of the boxes shown here are composed of two or three spaces, side by side. However, be sure to repeat the macro on every line of the menu that represents the box. That is, a triple-wide box will require three lines of the same menu item.

If you design and use a customized template, the screen menu will soon become superfluous. You can use it to consolidate what few commands are left over into a simple alphabetical list that should not occupy more than two screens. The pull-down menus can be programmed also. The author's pull-down menus are devoted to custom insertion routines for the entire library of blocks.

You will want a unique menu that reflects your own command usage routine. More than that, you will update it often as your

FIG. 9.1   DIGITIZER TEMPLATE

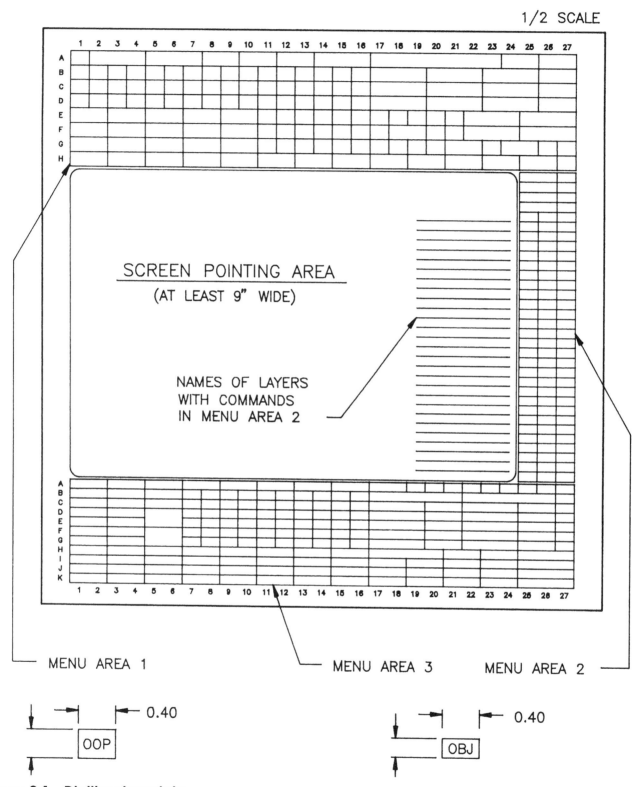

Figure 9.1   Digitizer template.

routine accelerates and you realize there are many possibilities for more shortcuts. The file that is used to draw the template can be color-coded by menu area, and the template can be plotted in color also. Of course, you will have to configure your tablet for the new menu. By designing and using a fully customized menu, you join the ranks of the AutoCAD power users and take a giant leap forward in your mastery of the program.

If your digitizer does not have a stiff, translucent plastic cover for the template, these covers are available at many CAD suppliers. Attach the cover to the digitizer with transparent, plastic tape that will form a hinge so you can lift the cover and enter revisions right on the template. An excellent tape for this purpose that will not fatigue and crack is made by 3M Corporation.[11]

Macro Descriptions

The menu items, sometimes called *macros*, shown in Fig. 9.2, the macro menu explanation, have been customized for designing plant layouts. The macros have also been designed to accelerate the process of issuing commands for AutoCAD. You can use them as is, or they can serve as a starting point for designing your own macros. If you are completely unfamiliar with designing macros, read the section about customizing AutoCAD in the appendix of your manual.

All 50 macros described in this chapter have been incorporated into an AutoCAD menu file named MACROS.MNU on the diskette. It can be loaded into the LAYOUT.DWG file quite easily by using the MENU command. Once loaded into your drawing, the macros will appear on two screen menus and can be executed just like any other menu item.

The macros are also available on the diskette in a DOS text file named MACROS.DOS. With a word processor set for DOS text, you can retrieve this file into the ACAD.MNU file or into your own menu file. In that way, you can then place the macros on your tablet menu, on the screen menu, or on the pull-down menus.

The value of selecting macros as an alternative to using the standard screen menu will become more obvious as you delve deeper into the process of designing them for your own use. To design a macro, just step through the commands that you want to include, and write down the symbol for every step that you perform. For an example, use any of the macros shown below.

If you want to try entering the macros manually into the standard AutoCAD template, start by loading ACAD.MNU into your word processor with the DOS text retrieve command. WordPerfect uses the TEXTIN/OUT command to retrieve a DOS text file. The tablet menu section is near the end of the ACAD.MNU file. In this section, right after the symbol [A1], type in any command that is shown in Fig. 9.2. Save the menu as a different DOS text file.

The explanation of each macro is shown just before the actual menu
item is described. This menu of line items can be retrieved into
your LAYOUT.DWG file with the MENU command. You can then test each
macro by picking it on the screen menu. Macro numbers A21 and A22
use the FILL layer. Macro numbers B12 and B13 use the NUMBERS
layer. These four macros will not work unless those layer names are
present in your AutoCAD drawing file.
```
  COPY WITH ORTHO ON:
    Copy a number of entities:
[COPY   +  A1 ]^C^CORTHO ON SNAP OFF COPY
    Copy one entity multiple times:
[COPY M1+  A2 ]^C^CORTHO ON SNAP OFF COPY \;M;^B
    Copy one entity once:
[COPY  1+  A3 ]^C^CORTHO ON SNAP OFF COPY SI;\^B
    Copy two entities once:
[COPY  2+  A4 ]^C^CORTHO ON SNAP OFF COPY \\;^B
    Copy three entities once:
[COPY  3+  A5 ]^C^CORTHO ON SNAP OFF COPY \\\;^B
    Copy four entities once:
[COPY  4+  A6 ]^C^CORTHO ON SNAP OFF COPY \\\\;^B
    Copy one window multiple times:
[COPY MW+  A7 ]^C^CORTHO ON SNAP OFF COPY W;\\;M;^B
    Copy one window once:
[COPY  W+  A8 ]^C^CORTHO ON SNAP OFF COPY W;\\;^B
    Copy last entity multiple times:
[COPY ML+  A9 ]^C^CORTHO ON SNAP ON COPY L;;M;
    Copy last entity once:
[COPY  L+  A10]^C^CORTHO ON SNAP ON COPY L;;
  COPY WITH ORTHO OFF:
    Copy a number of entities:
[COPY   x  A11]^C^CORTHO OFF SNAP OFF COPY
    Copy one entity multiple times:
[COPY M1x  A12]^C^CORTHO OFF SNAP OFF COPY \;M;^B
    Copy one entity once:
[COPY  1x  A13]^C^CORTHO OFF SNAP OFF COPY SI;\^B
    Copy two entities once:
[COPY  2x  A14]^C^CORTHO OFF SNAP OFF COPY \\;^B
    Copy three entities once:
[COPY  3x  A15]^C^CORTHO OFF SNAP OFF COPY \\\;^B
    Copy four entities once:
[COPY  4x  A16]^C^CORTHO OFF SNAP OFF COPY \\\\;^B
    Copy one window multiple times:
[COPY MWx  A17]^C^CORTHO OFF SNAP OFF COPY W;\\;M;^B
    Copy one window once:
[COPY  Wx  A18]^C^CORTHO OFF SNAP OFF COPY W;\\;^B
    Copy last entity multiple times:
[COPY MLx  A19]^C^CORTHO OFF SNAP ON COPY L;;M;
    Copy last entity once:
[COPY  Lx  A20]^C^CORTHO OFF SNAP ON COPY L;;
```

Figure 9.2   Macro menu explanation.

```
    OTHER MACROS:
    Form a window around a section of horizontal wall, hatch the
    wall and change the hatch to layer FILL:
[HATCH -    A21]^C^CSNAP OFF HATCH U;0;1;;W;\\;^B+
CHANGE L;;P;LA;FILL;;
    Form a window around a section of vertical wall, hatch the wall
    and change the hatch to layer FILL:
[HATCH |    A22]^C^CSNAP OFF HATCH U;90;1;;W;\\;^B+
CHANGE L;;P;LA;FILL;;
    Change an object to a different layer:
[CHG1 LAY   A23]^C^CSNAP OFF CHANGE \;P;LA;\;;^B
    Change the color of an object:
[CHG1 CLR   A24]^C^CSNAP OFF CHANGE \;P;C;\;;^B
    Change the elevation and thickness of an object:
[CHG1 E&T   A25]^C^CSNAP OFF CHANGE \;P;E;\T;\;;^B
    Erase one entity, repeated:
[ERASE 1S   B1 ]*^C^CSNAP OFF ERASE SI
    Change the line type of a line to hidden, no entry required:
[LINE HID   B2 ]^C^CSNAP OFF CHANGE \;P;LT;HIDDEN;;^B
    Change the line type of a line to continuous, no entry
    required:
[LINE CON   B3 ]^C^CSNAP OFF CHANGE \;P;LT;CONTINUOUS;;^B
    Make a trace of any width and length and then specify the
    elevation and thickness of that trace:
[TRACE ET   B4 ]^C^CORTHO ON SNAP ON TRACE \\\;CHANGE L;;P;E;\T;\;
    Rotate and move an existing object:
[R&M     1  B5 ]^C^CORTHO ON SNAP OFF ROTATE \;^B\\^OMOVE P;;
    Fillet two lines to a zero radius, repeated:
[FILLET 0   B6 ]*^C^CSNAP OFF FILLET R;0;;
    Select a line and break it, repeated:
[BRK LINE   B7 ]*^C^CSNAP OFF BREAK \F;^B
    Extend one line to a chosen point, repeated:
[EXT LINE   B8 ]*^C^CORTHO ON SNAP OFF CHANGE SI;\^B
    Change the wording of existing text:
[CH TXT S   B9 ]^C^CSNAP OFF CHANGE \;;;;;;^B
    Change the height of existing text:
[CH TXT H   B10]^C^CSNAP OFF CHANGE \;;;;\;;^B
    Enter text on center using the default font and size:
[DEF TXT-   B11]^C^CSNAP ON TEXT C;\;0;
    Enter a horizontal equipment number, change it to layer NUMBERS
    and move it, repeated:
[EQUIP# -  B12]*^C^CORTHO OFF SNAP ON TEXT C;\;0;\+
CHANGE L;;P;LA;NUMBERS;;^BMOVE L;;
    Enter a vertical equipment number, change it to layer NUMBERS
    and move it, repeated:
[EQUIP# |  B13]*^C^CORTHO OFF SNAP ON TEXT C;\;90;\+
CHANGE L;;P;LA;NUMBERS;;^BMOVE L;;
    Insert a block with no parameters:
[BLK NONE   B14]^C^CORTHO ON SNAP ON INSERT \\;;;
    Insert a block with rotation and move only:
[BLK  R&M   B15]^C^CORTHO ON SNAP ON INSERT \\;;\^OMOVE L;;
```

Figure 9.2 Macro menu explanation (Continued).

```
       Insert a block with rotation only:
[BLK    ROT    B16]^C^CORTHO ON SNAP ON INSERT \\;;
       Insert a right-hand three foot door with rotation:
[DOOR R36    B17]^C^CORTHO ON SNAP ON INSERT DR36 \;;
       Insert a left-hand three foot door with rotation:
[DOOR L36    B18]^C^CORTHO ON SNAP ON INSERT DL36 \;;
     View the drawing in 3-D from the right using the standard
     settings:
[3D RIGHT   B19]^C^CGRID OFF VPOINT 10,-8,9
     View the drawing in 3-D from the left using the standard
     settings:
[3D   LEFT   B20]^C^CGRID OFF VPOINT -10,-8,9
     View the drawing from the side:
[3D   SIDE   B21]^C^CGRID OFF VPOINT 1,0,0;ZOOM .8X;ZOOM W;
     Draw one line with ortho on, repeated:
[LINE 1S+   B22]*^C^CORTHO ON SNAP ON LINE \\;
     Draw one line with ORTHO off, repeated:
[LINE 1Sx   B23]*^C^CORTHO OFF SNAP ON LINE \\;
     Draw a rectangle with four lines:
[4 LINES    B24]^C^CORTHO ON SNAP ON LINE \\\\C;
     Set limits using OSNAP to find corners of the outline:
[LIMITS +  B25]^C^CSNAP OFF LIMITS INT \INT \^BGRID ON;
```

Figure 9.2   Macro menu explanation (Continued).

Return to AutoCAD, load your new menu, and, with the standard template configured on your digitizer, pick the "A1" box on the tablet and AutoCAD will lead you right through the macro. As you can see, entering macros is not difficult, and constructing them becomes routine with practice. One word of caution: It is not explained in the manual, but the SINGLE (SI) command will not work if any other letters besides control characters follow it in a macro. This fault, however, may be corrected in later releases.

Notice how the SNAP and ORTHO commands are closely controlled as each macro is executed. When you use these macros, they will seem to anticipate your needs and will become completely transparent to your thought process. That is the final test of a perfect macro: performance that anticipates without distraction.

In the examples shown, a label is placed at the beginning of each macro within a pair of brackets [ ]. If you use these macros on your screen menu, only the first eight characters within the brackets will be displayed. All characters within these brackets will be disregarded by AutoCAD during command execution. If you use the macro on your tablet menu, be sure to change the column-row symbol to identify the location of the macro on your template. Again, these macros were customized for designing plant layouts and can be easily changed to suit your more universal needs.

The first 20 macros shown in Fig. 9.2 correspond to the set of COPY commands that were mentioned earlier in this chapter. You can design a similar set of 14 MOVE commands from them by following the same general approach. As you get used to copying and moving entities with these tailored commands, you will see how easily the drawing process can be accelerated by using macros.

If you have not used macros or the MENU command before, here is the procedure for loading the macro list onto your screen menu, assuming you are using the LAYOUT.DWG file:

1. When you first load LAYOUT.DWG, the standard ACAD.MNU will be displayed on the screen menu and the blank layout drawing (Fig. 7.4) will appear on your screen drawing area.

2. Examine the layers shown in the layer dialogue box. They will be the same as those described in Fig. 8.1, settings for layout drawing.

3. Notice the layers named FILL and NUMBERS. They will be used by four of the macros when the selected entity is automatically changed to that layer at the end of the macro.

4. Type in the MENU command, and respond with the path and filename of the MACROS.MNU menu file. Do not add the extension .MNU to the filename.

5. AutoCAD will load the new menu file into your drawing, replacing the ACAD.MNU file. The screen menu will change and display the eight characters of the first 25 macros listed in Fig. 9.2, the macro menu explanation. The next 25 macros will appear when you pick item 2 on the screen menu #1.

6. This menu consists only of these 50 macros and nothing else. The program will not respond to any tablet or pull-down commands when the MACROS.MNU is active.

7. Study the explanations of the first few macros so you know what they are supposed to do when you execute them. Then try them out. The first 25 macros are tailored COPY commands and simple in function but very effective. The second 25 macros have some interesting combinations.

8. For instance, to use A21, draw a small horizontal rectangle, shaped like a wall, on the WALLS layer. Then pick the macro labeled "HATCH -" on the second screen menu. It will wait for you to put a window around the wall. Then it will hatch the wall with horizontal lines spaced at 1-in intervals and will automatically change the hatch entity to the FILL layer. Macro A22 labeled "HATCH |" on the second screen menu will hatch a wall vertically in the same way.

9. To use B12, pick the macro labeled "EQUIP# -" on the second screen menu. It will first ask you for the center point of the equipment number and then, after you pick it, the macro will ask for an equipment number. After you enter the number and hit Return, the number will appear and then change to the NUMBERS layer automatically. Then the number will become a phantom image, and the macro will ask you for a base point for moving. You can then pick a base and move the number into exact position. The macro will then ask you for the next number. The macro B13 will treat a vertical number in the same way.

10. Try all the macros, and then plan how you can customize your own screen and tablet menus for faster command execution. This planning strategy will continue for the rest of your career with AutoCAD, and your speed will accelerate with each new improvement.

10

Old Plant Layout

In this chapter, considerations for constructing large equipment blocks will be presented. The justification for drawing an old plant layout will be given along with directions for assigning number labels to the equipment blocks. Three illustrations of the Flexible Division's old plant layout will be shown.

Large Equipment Blocks

In Chapter 7, detailed instructions were given for using the library of blocks supplied on the diskette. Some guidance was also given for constructing blocks of unique equipment, and simplicity of construction was stressed. No mention was made, however, of how much time can be consumed in constructing blocks for large equipment.

If you have many pieces of large equipment in your plant, and many blocks to construct for them, your problem is twofold. One part is to make sure that all the new equipment that you draw is justified and should be included in the new layout. The other part is to construct large blocks that are simple and yet exhibit the functional appearance of the equipment.

Equipment Justification

The principle of avoiding symptomatic treatment of problems, which has been stressed in this book, applies even more as the final layout phase approaches. Large pieces of equipment are often purchased without any investigation of the actual need for that equipment. The equipment appears to be needed to solve a problem. In reality, the problem is hidden. The symptoms of the problem shout for attention and usually get it. Suppose, for instance, that this situation occurred on a golf cart production line:

Symptom: Two small body sections of sheet metal are supposed to lock together, but consistently separate and are unsightly.

Treatment of symptoms: Spot-weld the two pieces together, and smooth the joint. Another spot welder must be purchased and added to the line. The joint may fatigue and crack anyway. Next year's model will still have the same problem.

Problem: The two sections are bolted to two independent parts of the chassis, and the dimensional tolerance buildup prevents them from mating properly.

Cause: In the early design phase, the chassis was one structural piece, and then it was changed to two pieces.

Solution: Since it is too late to redesign the chassis, make one rigid part to replace the two sheet-metal parts.

By breaking the problem down into its components and directing your focus away from the symptoms, it will be easier to find the true cause. You can develop this technique of separating a problem into its parts with practice, by analyzing the evening news, for example. You will find that almost every "solution" to a social "problem" is actually a symptomatic treatment that will perpetuate the problem, not eliminate it. That is why we have the same social problems, over and over, forever. Do not let the same thing happen to your company.

Before you make blocks for new equipment that will be added to the line—in the previous case, a spot welder—investigate the actual need for that equipment. If there is a problem with the product, assume that it is only a symptom waving a flag over the cause, hidden deep within the process. Spend the money on sound product or process design instead of wasting it on more equipment to treat the symptoms.

Equipment Construction

A CAD program is capable of illustrating any apparatus in the finest detail, but it is not necessary to make each piece of major equipment a detailed work of art. On the contrary, simplicity should guide your equipment block design. Many of the great masters conveyed their impressions with a few simple brush strokes, and you can do it, too, in the layout field.

Every layout project is faced with a host of time constraints that cannot be ignored if the project has a critical completion date. Time saved in designing equipment blocks can be beneficially spent on the production floor, exposing and solving hidden problems related to the layout. An overly artistic layout with detailed equipment templates is soon forgotten. Process improvements, on the other hand, will have a lasting impact.

When you design blocks for large equipment, give your artistic talent a free reign. Look on the project as a challenge to convey the impression of the equipment by using the smallest number of traces possible. What small symbol can you add to the block that will show the viewer exactly what equipment it represents? Maybe a crank or a flag or a squirt of oil will convey the impression perfectly. In Fig. 12.2, the new plant equipment numbers, the IE used wavy lines on oven 707 and a starburst on laser trimmer 711.

Traces can be used at odd angles, too. Notice in Fig. 10.1, showing robot blocks in a 3-D view, the traces for the arms of the component insertion robots are shown at an angle, in the insertion mode. There are several more items to notice in this figure:

1. A gravity floor conveyor block from the library is used to indicate the powered, automatic conveyor of the assembly line.

2. The robot bases and other rectangular objects are traces.

3. The towers that support the arms are extruded decagons.

4. The towers have a circle on top to hide the internal lines.

5. The insertion pods are extruded decagons.

6. The production unit being assembled is shown mounted on the conveyor for more realism and clarity.

Do not use extruded circles to indicate round columns unless you want them to appear as black figures. AutoCAD automatically adds shading to extruded circles to better indicate the curved surface. This shading often becomes so dense that the column appears black and is overly prominent in a 3-D view. The shad-

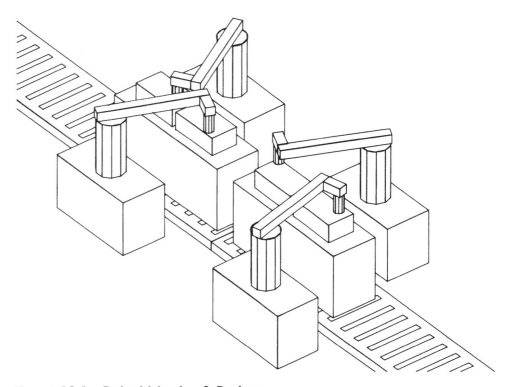

Figure 10.1 Robot blocks, 3-D view.

ing also expands the number of lines that must be hidden, which slows down the execution of the HIDE command.

Instead of extruded circles, use extruded decagons for drums or columnar sections. Extruded polygons do not have opaque tops, however, so add a circle to the top to form a cover if you want to hide the inner lines of the column.

You do not have to change the current elevation or thickness at any time while you are constructing blocks. It will be less confusing if you change each trace or entity individually. Also, leave the User Coordinate System at the default world setting and delete the icon. The robot components shown in Fig. 10.1 were easily constructed by entering the proper entity and changing the elevation and thickness by using the CHANGE LAST command. Here is the description of the robot components:

Component	Entity	Elevation, in	Thickness, in
Robot base	Trace	0	36
Robot tower	Decagon	36	36
Tower cover	Circle	72	0
Robot arm	Trace	72	6
Insertion pod	Decagon	60	12

An end view of large equipment is useful for coordinating the elevation and thickness of the component traces and other entities. Notice in Fig. 10.2, which shows robot blocks in an end view, how the top and bottom of each trace and extrusion are well delineated and how easy it is to check that these surfaces fit together.

You can create this view by entering the VPOINT command and responding with 1,0,0 (view from the right) or −1,0,0 (view from the left). The resulting side view image will be at the very bottom of your screen but can be lifted up with the ZOOM .9X command. Then use the ZOOM WINDOW command to move the image around on your screen. The PAN command cannot be used to move the image vertically because PAN is only effective in the XY plane (floor of the plant) of the drawing.

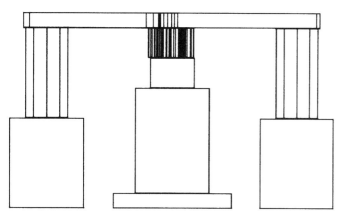

Figure 10.2   Robot blocks, end view.

It is best to use 6-in increments for the heights of all horizontal surfaces on your traces for large equipment. In that way, they will be easy to match up when you view them from the end. For instance, if you use five or six traces for a block, you can draw each trace, estimate the elevation and assign the proper thickness, then look at the end view and raise each entity up to the proper elevation. As you stack traces one on top of another, you will not have to memorize the heights of any group as you construct the block.

As you near the end of the construction, you can adjust both elevation and thickness by picking the entities while you are viewing the block from the end. You can alternate between the plan view and the end view to place each component in exactly the right location and to see that mating horizontal surfaces match up correctly.

You can create a macro that will ask only for the elevation and thickness of an entity by using the CHANGE command. The macro is shown in Fig. 9.2, the macro menu explanation, item A25. When you enter this macro and select the entity from the end view, you can watch it jump up and down and change thickness before your eyes. Again, Release 11 does not have this capability, but Release 12 does.

After you are familiar with the library of blocks on the diskette and have constructed all the blocks that are unique to your plant, you are ready to make a drawing of the existing floor plan. For the Flexible Division project, the IE called this drawing the *Old Plant Layout.*

Old Plant Layout Justification

The IE drew the old plant layout as the next step in the project schedule shown in Fig. 6.1. On the schedule, it is task 8, *IE: Draw old plant blocks and layout.* The necessity for drawing a layout of the old plant as it exists before the move is overlooked by many project leaders. At first, it may appear to be a waste of time, but upon further study, it actually saves time. By making an old plant layout you will be satisfying many steps at once, steps that would have to be taken individually anyway, often less efficiently.

Take the step of reviewing the new plant layout, for instance. Suppose that you have many benches in the assembly area of the new plant and you want to review your initial new plant layout with the supervisor. The supervisor will immediately ask you what function has been assigned to each bench in the new plant. Normally, there will not be enough room on the new plant layout for you to adequately label each bench. If you have assigned the same number to each bench on the old and new layouts, it will be easy for the supervisor to match the benches on the old layout with those on the new.

This system has proved very successful, even superior to labeling the benches on the new layout. The reason for this disparity is probably the supervisor's immediate understanding of the bench function in the old layout without having to interpret the designer's label. Because production supervisors are so familiar with the equipment in their area, they will instantly transfer the function of the bench on the old layout to the function of the bench on the new layout. As a result, the supervisor can quickly visualize how the benches will integrate with each other in the new layout.

Also, if the function of a bench on the new layout is different from its function on the old layout, the supervisor will see that change and know how much effort will be involved in making it successful. For these advantages alone, it is worth making an old plant layout drawing, but there are more. Here is a complete list:

1. *User review.* As already mentioned, the old plant layout can be used by the viewers to orient to the new plant layout. The old plant layout becomes a legend for all the equipment in the new plant layout as it is reviewed by the users, thus avoiding misunderstandings.

2. *Equipment numbers.* The old plant layout should be the basis for assigning equipment numbers to the existing equipment. If you made the equipment list shown in Fig. 5.1, then the old plant layout can be used to check off the existing equipment as it is recorded on the equipment list. Many errors can be eliminated by comparing the old plant layout with other documents.

3. *Equipment relocation.* Equipment will often be relocated from one supervisor in the old plant to a different supervisor in the new plant. The old plant layout can be used by both supervisors to locate and agree upon that equipment, thus eliminating a source of argument on moving day.

4. *Flow arrows.* The material flow can be indicated on the old plant layout and compared to the flow on the new plant layout. The flow improvement can thus be evaluated and improved further by everyone concerned. Management will also be aware of the improvement.

5. *Moving day preparation.* Before moving day, a self-stick label will be placed on each piece of equipment. The label will be color-coded to indicate the area where the equipment is to be placed in the new plant. The equipment number is also recorded on the label. If the old plant layout is used as a guide for placing these labels, it will ensure that all the equipment is recorded and numbered properly. Numerous mistakes can be avoided.

If you have not been given enough time to draw an old plant layout, you will probably make mistakes on the new plant layout and in planning for the move. If that is the case, inform management of all the problems that can be avoided by having an old plant layout, and urge them to allow time for it.

Old Plant Layout Preparation

The Flexible Division's old plant layout is shown in Fig. 10.3. Notice that every piece of equipment in the plant is included,

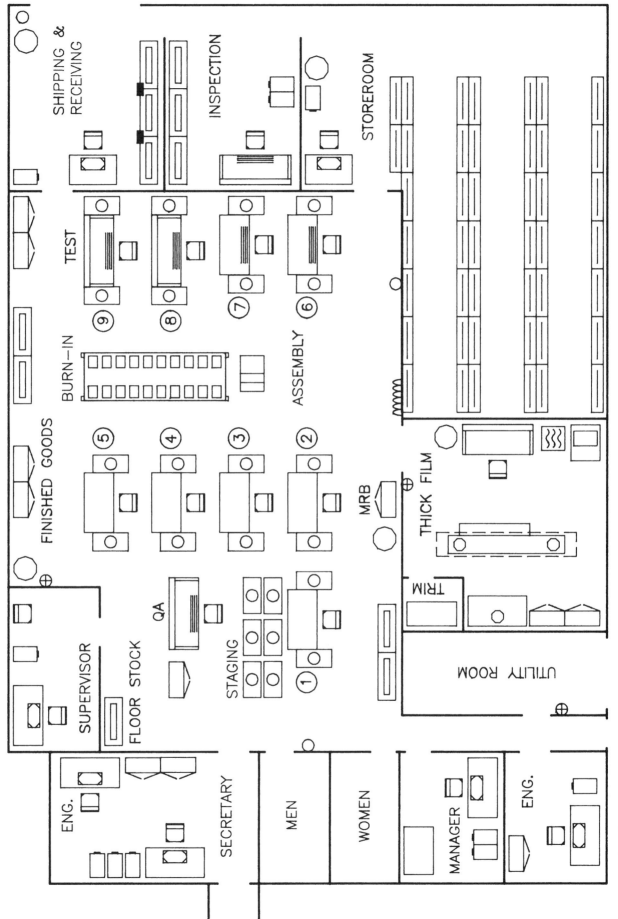

Figure 10.3　Old plant layout.

even the trash cans and fire extinguishers. Wastebaskets are the only exception. If they are important to your office layout, it is definitely overcrowded. Before constructing this drawing, the IE first gathered data concerning the dimensions and function of each piece. Follow these steps to gather this kind of data:

1. *Tour.* Before you start the old plant layout, tour the entire plant area that you will draw and become familiar with all the equipment. Notice any special utilities that will have to be disconnected before the equipment can be moved. Notice any large equipment that will have to be disassembled before it is moved. Notice if any internal walls will have to be opened up before the equipment can be removed from a room. Generally familiarize yourself with all the problems that will be encountered before moving day, and decide how to solve them. If you discover a problem that you can't settle, discuss it with the people concerned and form a plan of action.

2. *Sketch.* Divide the old plant layout area into sections that can be drawn easily, and notice what equipment is in the first area to be inventoried. Make a rough sketch of the walls, door openings, closets, etc., not to scale but reasonably proportioned. Add a sketch of each piece of equipment; in most cases, just the rectangular outline will do. For large equipment, make a detailed sketch for the group of traces or other entities that will be used.

3. *Dimensions.* If you made an equipment list as shown in Fig. 5.1, then you have already measured the equipment and assigned numbers to each piece. If you did not make the list, measure the equipment and write the width, depth, and height next to each piece. For large equipment, enter the dimensions and elevation of each entity on a separate sheet of paper. Save all paperwork in your notebook.

4. *New Equipment.* If any equipment that will be used in the new plant is not in the old plant, that equipment must be measured and sketched also. The equipment may be in a warehouse or may be still on order. Always measure the actual equipment if possible; otherwise, use the dimensions on the specification sheets. Construct blocks for the new equipment, and prepare them for the old plant layout just like the old equipment. As you finish drawing the equipment in each area and it is fresh in your mind, consult with the users about its function.

5. *Equipment function.* If you have made the process flowcharts as shown in Chapter 2, you are very familiar with the main function of each bench and other assembly equipment in the production areas. But that is not enough. This is the time to learn the function of every item in every room that you will lay out. That includes the trash cans, cabinets, shelves, and other items in the assembly areas. That includes fire extinguishers, eye wash stations, first aid stations, supplemental oxygen, nurse's room, and other health-related equipment. That includes all the furniture and storage equipment in the offices. Question the area personnel, and gather enough information about each piece of equipment so you know why it is located where it is and whether that location can be improved.

6. *Final data check.* Review all your sketches and other data for any items that you may have missed. Take instant pictures of very complicated equipment or complex equipment groups. Tour the areas once again, in greater depth this time, to be certain that no item has escaped your survey. Look behind the large equipment for hidden pumps, floor drains, drip pans, trash cans, or clearance areas. These clearance areas must be added to the equipment template. Find out why the clearance is required, and add a dotted line and dimensions to the equipment sketch to indicate the extra area. Make your sketches as neat and complete as possible, so while you are drawing the old plant layout on your computer, you will not have to return for missed data.

Old Plant Layout Construction

After gathering the required data, the IE made the Flexible Division old plant layout shown in Fig. 10.3. The walls and other structures are not to scale, but the proportions are fairly accurate. The IE used the library of equipment blocks to represent most of the equipment. Follow these steps as you construct the old plant layout drawing:

1. *Drawing file.* Retrieve the file LAYOUT.DWG, and use the directions given in Chapter 8 to prepare it for your old plant layout drawing. Erase the example layout blocks, but retain the block construction area. Select an outline that can be used at the $\frac{1}{8}$ in = 1 ft scale, and erase the others. You should be able to make the entire old plant layout within the outline that you have chosen.

2. *Walls.* Draw the walls on layer 1 with a single line, but plot them with a wide pen to make them more prominent. Do not bother to measure the rooms, but make the dimensions and proportions reasonably close to the actual room dimensions. Do not include any doors or windows, but leave an opening where the doors are.

3. *Standard equipment.* Place all the equipment for each work center on a different layer with a different color. The easiest way to insert equipment blocks from the library is to locate the name of the equipment in Fig. 5.1, the new plant equipment list. Then use the INSERT command, type in the name, and place the equipment where you want it. After that, you can copy it to other locations.

4. *Unique equipment.* Use view B, the block construction area of the drawing, to construct the equipment that is not in the library. Use view C, the 3-D view, to see how your block will appear in three dimensions. Use view F, the side view of the block construction area, to check the matching horizontal surfaces of the entities in the block. After the block is complete, assign a name to it that will conform to the names in the library, as described in Chapter 7, and insert the block into the old plant layout drawing.

5. *New equipment.*   Use view B to construct the blocks for new equipment. After they are complete, make them into blocks and insert them next to the old plant layout out in the margin where they will be plotted along with the existing equipment.

6. *Layout labels.*   The old plant layout labels, such as ASSEMBLY, should be placed on a separate layer so they can be turned off as a group. Each work center should be identified along with the offices, rest rooms, utility room, and other rooms that have a specialized function. Also, identify any cabinets or shelving that are used for a special purpose such as finished goods, first aid, Material Review Board (MRB), and floor stock. Notice the various labels shown in Fig. 10.3, the old plant layout.

Identify the operation that is performed at each assembly workbench with a circled operation number. The numbers shown in Fig. 10.3 correspond to the numbers that were assigned to the operations in Fig. 4.1, the old CEDAR Labor Analysis.

Turn off the layer that you used for the labels, and place the equipment numbers on a new layer. Place the numbers right on top of each piece of equipment if possible; if not, right next to it. The numbers may be on your equipment list already, or you may assign them directly on this drawing. Assign a different number series to each work center, as shown in Fig. 10.4, which shows the old plant equipment numbers.

This number series and the color-coded labels will be used on moving day to place the equipment in the proper location according to the new plant layout. They will be used constantly by everyone involved in the move, so keep the numbering system simple. Assign only three digits to each series. Notice that even the fire extinguisher and trash cans are numbered. Although wastebaskets are not shown on the layout, they should be labeled with the same number as the equipment they go with. This may seem like overkill right now, but on moving day it will pay off in less confusion and fewer unanswered questions.

Notice the shelf racks in the upper right-hand corner of Fig. 10.4, which shows the old plant equipment numbers. The equipment numbers are 601, 602, and 603. The black rectangles between the racks indicate that they have common end supports. If these racks are separated in the new plant, more end supports will have to be purchased.

The AutoCAD ATTRIBUTES command can be used to assign data to the equipment and analyze it, but this method is very time consuming. If you want to constantly update the layout and use it to accumulate permanent information, then use a facilities management program. There are more than 50 programs that offer a seamless link between AutoCAD drawings and the database. A listing is shown in *The AutoCAD Resource Guide,*[3] under Facilities Planning and Management. This guide is normally included in the AutoCAD program package. Here is a small sample of the services that most AutoCAD facilities management programs will provide:

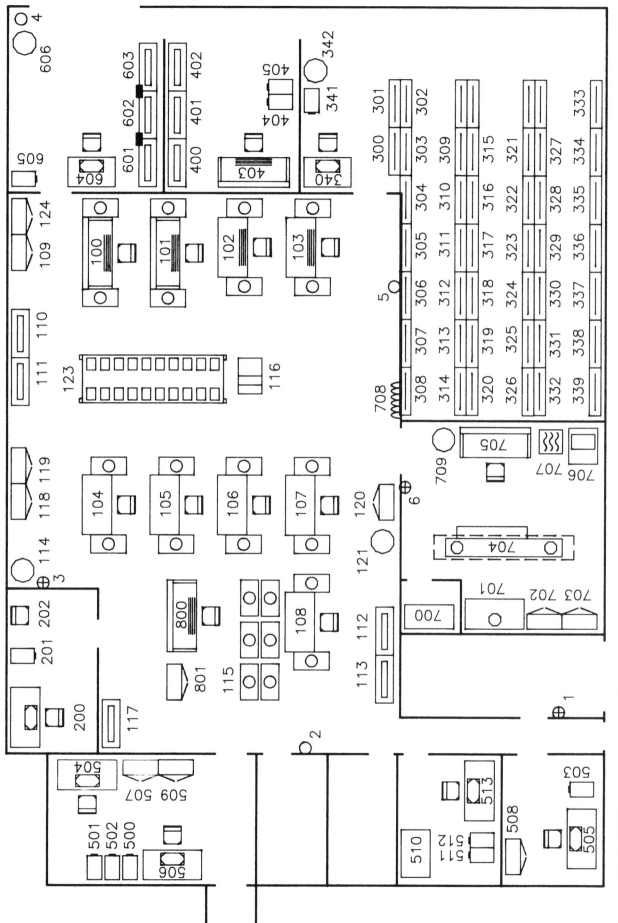

Figure 10.4   Old plant equipment numbers.

1. *Equipment numbers.* You can assign and record equipment numbers, then move and rotate them independently of the equipment.

2. *Equipment blocks.* You can use any AutoCAD library of blocks and other 3-D blocks that you design.

3. *Equipment location.* It will highlight a piece or category of equipment on the layout after you supply the proper number.

4. *Data.* It will accumulate a wealth of data on equipment inventory, cost, departmental areas, utilities, space management, and much more.

5. *Communications.* It will automatically draw lines on your layout that link computers to networks, telephones to switchboards, and many other utilities of this nature.

6. *Maintenance log.* You can record maintenance information on all equipment, and it will signal when maintenance is due.

7. *Reports.* It will generate standard reports for all activities related to facilities management.

Old Plant Layout Material Flow

The material flow on both old and new plant layout drawings is indicated by a set of flow arrows named FLOW-xx in the block library. The flow in the Flexible Division's old plant layout is shown in Fig. 10.5.

The memory bank products for the Flexible Division were transported in carts and assembled at benches that were generally placed in sequence. The flow was haphazard and caused numerous errors in assembly and test. The status of any one order was almost impossible to determine, even in this small shop. That was the reason for the daily status and priority meetings. The flow arrows illustrate this condition very quickly, just the form of presentation that management appreciates.

To use the most common flow arrow, insert the block named FLOW-ST. The base point is at the end of the shaft, so the arrow can be placed and then rotated in the direction of the flow. You can make a macro that will repeat this insertion forever, until you interrupt it. With it, you can place a whole string of arrows very quickly. The macro looks like this:

```
*^C^CORTHO ON SNAP ON INSERT FLOW-ST \;;
```

When you issue this command, a phantom image of the flow arrow will appear. With SNAP and ORTHO on, place the shaft where you choose, rotate it, and release it, and another phantom image will immediately appear, ready for the next placement.

When your old plant layout is complete, review it with the area supervisors to make sure that all equipment is accounted

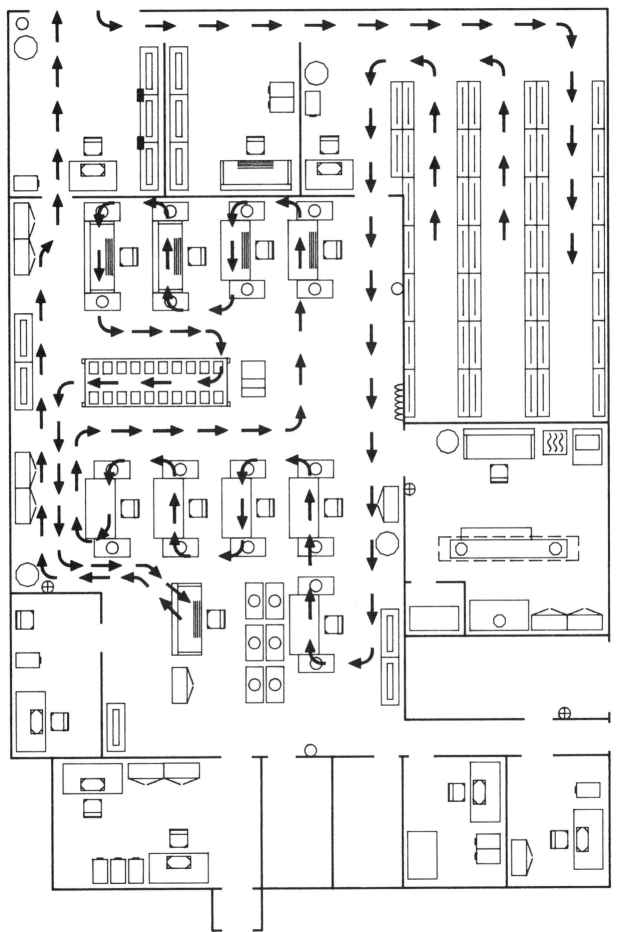

Figure 10.5   Old plant material flow.

for. All of the equipment that will be included on the new plant layout should be shown on the old plant layout drawing for these categories:

1. The equipment present in this area right now.
2. The equipment in some other storage area.
3. The equipment in some other work center area.
4. The equipment that will be purchased.

11

New Plant Layout Preparation

In this chapter, various elements that you should consider before starting the new plant layout will be explained. A more detailed explanation of the improvements checklist will be given, and three more lists will be presented: the errors checklist, the omissions checklist, and the design tips checklist.

Plant Layout Goals

What do you expect to accomplish with the new plant layout? Is it possible that the plant layout project is premature and should be postponed until the present manufacturing system is more refined? To answer these questions and more, consider these points when you set goals for your new plant layout:

1. *Product cost.* Reducing the product cost is often mentioned as a goal for a new plant layout. However, unless the present plant operation is chaotic, it will be almost impossible to reduce the product cost significantly only by designing a new layout. Even though the old layout may be better than the new one, they will both perform the same function and result in essentially the same product cost.

The labor data shown in Fig. 4.1, the old CEDAR Labor Analysis, can be used to compare the labor costs of a new layout with those of an old layout. Notice the components of the total labor hours, and how little effect a new layout will have on them if no changes are made in the system. If you want to lower the product cost, first improve the manufacturing system by attacking all components of the product cost, then change the layout to implement those improvements.

If you have constructed the spreadsheet to estimate your total labor hours, then it can also be used to justify or reject a change in the new layout. Create a before-and-after spreadsheet to compare the labor hours for the old versus new layouts. Then decide if the minor changes you have made actually save labor, and more importantly, notice what changes you can make that will reduce the product cost in other ways.

For high-tech products, the labor is a minor part of the product cost, so the IE devoted little time to improving it. To improve your product cost, concentrate on all means available before you start the new plant layout. Simply moving to a new location without improving the product cost is the ultimate example of a lost opportunity.

2. *Production expansion.* Increasing the output capacity of the manufacturing area may be the most common goal for a new layout. In addition to expanding the assembly area, this increase may affect all production service groups, from purchasing to sales. Each work center must be studied for capacity-related functions, including the shipping dock, the receiving dock, and the storeroom. The details of expanding a storeroom will be explained in the next section.

Figure 4.1, the old CEDAR Labor Analysis, and Fig. 4.2, the new CEDAR Labor Analysis, can be used to calculate the additional work stations that will be needed to expand the capacity of the assembly area. Change the operations and the minutes-per-unit values in Fig. 4.1 so they represent your procedure exactly as it exists before the expansion takes place. The summary figures for the pie chart data will represent your system before it is changed. Since the system is being expanded and not changed in any other way, this summary can be compared to the new summary on Fig. 4.2 to verify that the system is still just as efficient as it was before the expansion.

Naturally, you may have to make major revisions in this spreadsheet before it becomes an accurate mathematical model of your process. If you have constructed a new process flow-chart like Fig. 2.3 for your system, it will help you avoid any omissions.

Next, change Fig. 4.2 to represent your process after the expansion has taken place. Adjust the figures upward under the column heading NUMBER PEOPLE, STATIONS. As you do, the number in the upper section, UNITS PER WEEK, will increase also, indicating the new capacity.

Continue to adjust the NUMBER PEOPLE, STATIONS figures upward while you watch the output approach your goal. Let the bottleneck indicator be your guide while you consider adding people, work stations, overtime, and other shift hours that will be needed to achieve the desired output. In the end, you should arrive at the desired capacity and have no obvious bottleneck in the system. The number of people, stations, and other equipment that you have entered will represent what is required to achieve the desired output.

While you are adjusting the main body of the spreadsheet, the summary section will reflect any changes in the flow time and efficiency of the system. Examine this section for accuracy. If the flow time or efficiency has deteriorated, investigate the cause.

3. *Travel distance.* Reducing the material travel distance between departments is another goal that is often set for a new layout. It may be proposed that a large section of the plant be moved, at enormous cost, merely to achieve the close proximity of two departments.

Evaluate these projects closely for economic justification and flow time reduction. Often, the material is transported in a conveyance that contains a large number of parts, and the dollar savings involved in reducing the distance are minimal. Divide the labor dollars involved in the movement by the number of units being moved. Notice the very small effect that these savings have on the end product cost. Then consider what other, more cost-effective improvements could be made with that same dollar expenditure.

Not only that, if the operators can be rotated for this job and everyone takes a turn, they will look forward to a chance to get up and move around occasionally. If that working break is not available, the operators may figure out some other way to move around on their own.

Do not let short travel distances by themselves dictate your design strategy. Not that short travel distances are unimportant; they can be very important, but there are also other cost considerations.

4. *Flow.* Reducing the flow time should be the dominant goal when you are designing a new plant layout. Improving the flow of the manufacturing system can affect the operation of the entire plant, from purchasing to sales. The ripple effect on everyone's job can be impressive. All employees, including the department managers, should investigate their work procedures for ways to improve the flow time.

Before you start the mechanics of designing the new layout, concentrate on improving the flow time of the manufacturing system. Many times, it can be shortened by simple method improvements, such as changing to smaller batch sizes or kitting the parts for each assembly operation. If you can make significant improvements in the flow, question the need for a finished-goods inventory.

While improving the flow, you may encounter symptoms of hidden problems. For instance, excessive rework can increase the flow time substantially. You can treat the symptom by designing better tooling for rework (a red flag should go up in your mind), or you can solve the problem by establishing 4Q vendors for all critical parts, qualifying the operators, and qualifying the processes. Rework is one of the principal targets for flow time improvement.

Unless you make significant changes in the flow time, the new layout by itself will not do it for you. Concentrate on each individual step in the manufacturing process for excess flow time. Reduce the queue hours shown in the old CEDAR Labor Analysis, Fig. 4.1, that apply to your process. Improving the product flow time is an ongoing effort involving everyone, and enlisting their help should be one of your goals.

5. *Workers with disabilities.* A new plant layout may be needed to accommodate employees with disabilities. In 1990, the Americans with Disabilities Act (ADA) was signed into law. A disability is any mental or physical impairment, except for those caused by the use of illegal drugs or alcohol, which imposes a limit on the function of an individual. All employers are required to comply with the regulations and provide "reasonable accommodations" without "undue financial hardship." It is beyond the scope of this book to describe all the various changes that should be made to the mechanical structure of a facility to accommodate a person with a disability.

There are, however, many points directly related to the design of a plant layout that can provide for "reasonable accommodation." Whether your company now employs workers with disabilities should not affect your approach to these considerations. All facilities should be designed so that hiring and accommodating workers with disabilities will be a matter of routine. The dimensions given in this book for aisle widths, rest room facilities, workbench clearances, and other measurements generally conform to the ADA requirements.

At this point, if you do not have a qualified person, an architect, for example, on the plant layout project team, you should plan on who will advise you. It is far easier to make an alteration in your plant along with the layout than to retrofit it later at a much higher cost and greater inconvenience. More information on this subject will be provided in Chapter 14.

6. *Supervision.* This is seldom mentioned as a goal of the new plant layout, but in many ways, supervising an area can be easy or hard, depending upon how it is laid out. These items make it difficult for a supervisor to control an area efficiently:

   a. The area is fragmented into many subdivisions.
   b. The area is very long and narrow.
   c. The area is distant from the supervisor's desk.
   d. The area has many rooms containing operations that cannot be easily observed from the outside.
   e. The material flow is difficult to follow, and operators must be trained to expedite some units and not others.

During the review process, the supervisor may not notice that these difficulties exist. If these awkward arrangements are incorporated in the layout and become apparent only after moving day, it will place an undue burden on the supervisor. Advise the supervisor of potential problems while you are both reviewing the layout.

Other Plant Layout Considerations

Review these subjects, and if they apply to your plant, become more knowledgeable in their application to your plant layout design:

1. *Electrostatic discharge (ESD).* If your product is sensitive to ESD, then proper precautions have probably been established in the assembly area. For more information about the importance of ESD and how it may apply to your product, see *Electrostatic Discharge Control.*[14]

The fact that an area produces products that are sensitive to ESD will have a minimal effect on the layout. The equipment and other items needed will be more employee- and work station-oriented than layout-oriented. The employee- and work station-oriented items will include:

a. Conductive smocks

b. Conductive wrist straps

c. Conductive bench mats

d. Conductive bags

e. Antistatic unit holders

f. Conductive shoe straps

g. Ionizers

h. Conductive carts

i. Antistatic tote boxes

The layout-oriented items include the following:

a. *Conductive floor tiles.* These are connected to the ESD ground system of the plant. The tiles can be specified on the electrical utility overlay of the layout.

b. *Conductive floor mats.* They are used in place of tiles at selected work stations. Make a symbol block for the bench to indicate that a floor mat is under the bench.

c. *Conductive benches and chairs.* These are relatively expensive and should be used only on those operations that are ESD-sensitive. There are symbol blocks in the library named ESDHIGH and ESDLOW that can be used to indicate conductive benches.

d. *Ground-fault indicator receptacles.* They are designed to ensure that the 110-V supply in the plant is properly grounded. Also, if hand tools used at the work station are inadvertently grounded, the relay in the receptacle will turn off the power. Since the operators may be grounded in an ESD work station, they are especially at risk to severe shocks. These special receptacles can be specified on the electrical utility overlay of the layout.

e. *ESD ground system.* It is isolated from the electrical ground and the computer ground systems of the plant. It is used to provide a neutral ground for the operators and nonelectrical components of the ESD system. This system can be indicated on the electrical utility overlay of the layout.

f. *Humidity control.* Since static electricity can be generated more easily in low-humidity conditions, humidity control in an ESD-sensitive area is essential. The ideal humidity varies with the product and is usually controlled within the ventilation system. The specifications can be shown in a note on the mechanical utility overlay of the layout.

2. *Electromagnetic fields.* Yet another hazard of modern technology has surfaced recently, that of 60-Hz electromagnetic radiation. This radiation is present, to a greater or lesser extent, in every electric conductor around us, and the degree of the exposure varies with the duration, current intensity, and proximity to the source. The intensity of the exposure, as with all radiation, varies as the square of the distance from the source.

Although everyone will be affected to some extent by exposure to this radiation, children and pregnant women are especially at risk. The book entitled *Warning: The Electricity Around You May Be Hazardous to Your Health*[10] presents a history of case studies measuring the effects of magnetic radiation on humans in the 60-Hz frequency range, called *extremely low frequency.* It is a convincing chronicle of the accumulation of knowledge indicating that electromagnetic radiation can promote cancer in humans and their unborn children under certain conditions.

Electromagnetic radiation also has the beneficial effect of promoting bone growth after a fracture. Unfortunately, it promotes all cell growth without preference, including cancer cells, so it cannot be used on patients with tumors. Although electromagnetic radiation is not a carcinogen, it acts as a promoter to a precancerous growth in the body.

Electromagnetic radiation may or may not influence your layout design, depending upon the equipment in your plant. A small appliance held close to the body may be just as hazardous as a huge dynamo 10 feet away. You have to measure the field intensity with a Gauss meter to be sure. The manufacturers of these meters are listed in *Warning*[10] along with instructions for measuring magnetic fields. You can buy a meter as a permanent reference tool or rent one; or your electrical contractor may have one.

The safe level of radiation for prolonged exposure is still being debated, but is probably between 1 and 3 milligauss (mG). Sources of exposure can be any high-current machinery, power lines, or any electric device that is held near any part of the body. If possible, discuss the ambient magnetic field conditions with your electrical contractor before the plant wiring is installed in the new facility.

The best strategy to avoid undue exposure is to become knowledgeable on the subject, buy a Gauss meter, and measure the radiation of any suspected source in the plant. If you discover a high-radiation area around some equipment or in the new plant walls, restrict the traffic around that source. Do not assign a work station to that area where someone may sit all day.

If you buy new equipment, ask the manufacturer if electromagnetic radiation has been reduced to a safe level. Try to have hazardous machinery rewired for low magnetic field intensity, or install a steel shield around it.

Measure the radiation around all computers, as the level can vary significantly. On your layout, do not assign permanent seating to anyone within 4 feet behind or to the side of a computer unless you have verified that the area is safe. Generally, computer operators are safe if their eyes are 2 feet away from the monitor, but measure the radiation there, too. Old computers may have to be replaced.

As the industry becomes more aware of this problem, more plants and equipment will be designed to eliminate it. Until then, recognize that magnetic radiation can be a hazard to you and the people in your plant. Then design your plant layout to minimize that risk.

3. *Noise control.* Ambient low-level noise can be an insidious problem. Ask the people in a computer room, for instance, if they are bothered by the constant hum that comes from a large equipment cooling fan. Invariably they will say, "No, I'm used to it." Then, during a power failure when the fan is off, ask them again. A common response is, "I wish it were like this all the time."

That's the typical low-level noise situation throughout most plants, and you have the opportunity to improve it in your new plant layout. Often, this low-level noise can be attenuated by using a simple barrier composed of three office partitions placed around the offending equipment.

If the equipment noise is somewhat louder than that, a simple enclosure will kill it completely. Use ¾-in plywood for the box, and line it with acoustical foam. Your acoustical foam dealer can recommend what configuration will attenuate the source frequency. Be sure to vent the box from end to end if cooling is a problem. Isolate the vibration of the equipment by placing it on a special pad, specified by your dealer. Make the pad large enough to extend out to the inside edge of the enclosure. Many more practical solutions can be found in *Noise Control, A Guide for Workers and Employers.*[15]

As the noise becomes more intense, loud enough to cause damage, the solutions must be more sophisticated. High-intensity noise abatement is beyond the scope of this book, but a good reference is *Industrial Noise Control.*[13] It contains some practical solutions and conforms to OSHA regulations. Also, noise measurement and the effect of noise on humans are covered in *Handbook of Industrial Engineering*[1] (p. 1047).

If the noise level in your plant borders on the uncomfortable range, then your company should invest in a sound level meter. It combines the magnitude of the sound for all frequencies into one number called the *noise level.* Measure the noise level in those areas where it may be a problem, and decide whether any structure or new equipment will be needed to reduce it. Then, if it affects your floor plan, incorporate those improvements into your new plant layout.

4. *Storeroom expansion.* When you design a new layout for expanding the production capacity of the assembly area, the storeroom may or may not have to be expanded also. If all critical parts are being supplied by 4Q vendors and therefore can be

delivered more frequently, no adjustment may be necessary. However, the stock levels of all parts should still be analyzed for problem areas.

Divide the stock into common groups according to the vendors or the type of storage required or the packaging containers, whichever applies to your situation. Then decide how each group can be adjusted to respond to an elevated demand. If you feel that the storage capacity must be expanded for one group, it will probably not be linear. As an example, only half again as much shelving may be needed to support a doubling of the production area. In most cases, a mutual decision by you and the storeroom supervisor will be adequate for moderate expansion situations.

If you are starting out from scratch without the advantage of an existing storeroom, the problem is much more complex. Again, divide the parts into common storage categories, for instance, floor storage, pallet racks, bulk shelving, small-parts shelving, and floor stock. Decide what sequence the parts will be stored in, stock number sequence or kit assembly sequence.

Do not discard kit assembly sequence too hastily because it first appears to be disorganized. Actually, part locations are usually memorized by part appearance rather than by stock number anyway. The association of part to bin is learned very quickly by the stocking personnel. More time is spent pulling stock than replenishing it, and the pulling is far more efficient when parts are in kit assembly sequence. That is, every part in each category is located on the shelf in the exact sequence that it will be pulled to load the issuing container. Parts that are common to several containers are stored in a separate area.

After the sequence decision is made, calculate the actual bulk storage volume of the inventory for each category. Then multiply that volume by an expansion factor corresponding to the type of storage equipment. This factor is an estimated value that compensates for the unused volume of the shelving units. You should estimate this factor for your application. It will be about $2\frac{1}{2}$ in most cases. That is, multiply the bulk volume of the parts by $2\frac{1}{2}$, and divide the result by the actual storage volume of the storage unit. For example,

$$\text{Bulk volume of small-parts boxes} = 80 \text{ cu ft}$$

$$\text{Expansion factor for light shelving} = 2\frac{1}{2}$$

$$\text{Usable volume of a shelf unit:}$$

$$4 \text{ ft wide} \times 5 \text{ ft high} \times 1 \text{ ft deep} = 20 \text{ cu ft}$$

$$\text{Number of shelf units required:}$$

$$\frac{80 \times 2.5}{20} = 10 \text{ shelf units}$$

The final figure will be the number of storage units that you will need to store each category of stock. If you believe the shelving figure is too tight or you want a better safety factor, use a higher number for the expansion factor.

5. *Clean room.* Although clean-room construction is a specialized field not covered in this book, a clean-room floor plan is generally no different from that of any other production room. However, the equipment within the clean room will probably be unique, such as disk format machines or wafer processing equipment, and you will have to construct these blocks. Consider these clean-room guidelines for your layout:

a. Compared to the production area, allow more space for the equipment and the clean-room personnel. The room itself can limit the occupants' freedom, but the volume-to-people ratio can also affect the actual class of the clean room. A class 10 room can quickly become a class 100 room once the operators arrive. More room volume will help that situation.

b. Allow excess area around the equipment for maintenance personnel in full clean-room clothing.

c. Allow excess area in the change room also. Assume that everyone will suit up at once, and size the room accordingly.

d. The change room can be considered a gray area in some applications and can be used to house intermediate, gray-room equipment. A laser trimmer, for instance, can be accessible to microcircuit clean-room operators and normal-dress engineers at the same time.

e. The supervisor's office can be built as an intrusion with three sides exposed to the clean room, but the fourth side open to the plant. Install pass-thrus and talk-thrus in the dividing walls for communication.

List of Improvements

Figure 3.1, a list of improvements, can be used as a starting point to make your own list and expand your knowledge base. The resulting guide will directly affect the layout design and should be referred to and updated often as the layout proceeds. If you do not improve the assembly procedure, all the effort that you expend on the layout will have no significant impact on the product cost or flow.

An improvements list can also lead to superior product quality. The changes that you make just to smooth the product flow can result in fewer assembly errors, less rework, and a final product with fewer inherent flaws.

If you plan to make a new plant layout presentation to management, the improvements list can be a valuable guide and an impressive handout during the meeting. As discussed in Chapter 9, however, the meeting will be justified only if the interaction of the participants is important. Otherwise, if the layout has been previously reviewed and will be readily accepted, just distribute a memo along with the improvements list to everyone concerned.

As part of your layout preparation plan, query the users for more possible improvements to add to your list. Use a one-on-one interview to obtain a more candid response, and avoid holding a meeting with all the direct labor personnel in one room. Make suggestions, elicit opinions, and generally draw each person into the program by making sure that she or he participates in the layout design. As you leave each interview, keep the line of communication open by encouraging the person to stop you at any time with any light-bulb ideas.

Errors, Omissions, and Tips Checklists

All the lists described in this section should be referred to often during the layout design phase. If you discover more items that are applicable to your plant or industry, add them to the lists for future reference. Again, these lists also serve to expand your knowledge of the industry.

Figure 11.1, the errors checklist, should be used as a reminder of the many errors that can be made during the layout process. Even if you have conducted other layout projects, error avoidance must be one of your top priorities. In the beginning, this may not impress you as being a major goal, but on moving day, if you discover a major error that you have made, the results can be dramatic and long-lasting. You may have specified a doorway, for instance, that is too small for the equipment that must be moved through it. You will be reminded of that one for years to come.

While you are reviewing the layout drawing with the users, ask them about any details that have error potential. It will be very easy for you to commit a major error in the layout when you are concentrating on the placement of equipment and not contemplating the details.

Figure 11.2, the omissions checklist, should be used as a reminder of the many omissions that may be made during the layout process. Omissions differ from errors in that they relate to features so common that they may be forgotten by even the most experienced layout designer. These situations can be embarrassing. Suppose you present the general contractor with your finished layout, the result of three months' work, and the contractor asks you, "Where are the fire extinguishers?"

One common omission in any layout is a first aid station. The supplies in the station should be tailored to the type of injuries that the employees are exposed to. The IE for the Flexible Division converted one of the finished-goods cabinets into a first aid supply center, since the cabinet was no longer needed for finished goods. Because of the generous size of the cabinet, many items were included that are not normally present. A collapsible stretcher, a respirator, and a neck brace are just a few.

Another common omission is an eyewash station. Notice that the IE made room for it right next to the drinking fountain, with

Review the following list of errors before you design your new plant layout, and look for other items in your project that have error potential. When you find them, add them to this list for future reference. Opportunities for error are endless in plant layout projects, and error avoidance is a primary concern.

1. The architectural drawing dimensions were used to make the new plant layout drawing instead of measuring the new plant. The equipment will not fit into the space provided because the actual plant wall locations do not agree with the architectural drawing. Moving day is a catastrophe.

2. The door opening of a room is not large enough to accommodate the equipment that must go through it. After moving day, the old door and frame must be removed and a new, wider door installed.

3. The columns in the building were not shown on the layout drawing and thus interfere with the equipment placement on moving day.

4. The location of the exhaust duct in the ceiling for the bench hood was not specified accurately. When the bench is moved into place, it is misaligned with the bench exhaust vent. The contractor has to spend extra time and money to route the ductwork properly, and, even then, the misaligned duct is unsightly.

5. Electrical panel clearance for the electrician was not shown on the layout and is not taped off. (This is not necessary if the panel adjoins an aisle.) On moving day, the equipment that interferes with that area must be relocated.

6. Structural intrusions such as heating ducts near the ceiling were not shown on the layout and thus interfere with the pallet racks on moving day.

7. Double doors were not specified leading to the shipping or receiving areas for moving large items in or out of the plant. The single door must be removed and the wall opened up on moving day.

8. The incoming inspection area is not walled off and secure. As a result, parts can easily be extracted for production use before they are processed.

9. The supervisor's office is too far from the area of supervision, creating a lack of routine communication.

10. There is not enough access room around the gravity conveyors. If parts get hung up, they cannot be freed easily.

11. The new plant overall measurements were not double-checked, and major errors were made. As a result, one whole area in the new plant is overcrowded while the adjacent area has more than enough room.

12. Drip pans were not provided for machinery that by its very nature emits oil or lubricants. The oil around these machines is a safety hazard and requires excessive cleanup time. A drip pan should have been discussed with the operator and designed before moving day.

13. The actual width of some unique pallet racks was not measured before the blocks were constructed. The error was cumulative, and on moving day a long row of connected pallet racks extends into the aisle. One of them must be removed, and its storage capacity is lost. In most

Figure 11.1   Errors checklist.

cases, the nominal width of a pallet rack (8, 10, 12 ft) does not include the end support posts. That is, the actual width of a 10-ft pallet rack in a row of racks is 10 ft 3 in. (The width of the pallet rack blocks in the library has been extended to include the posts.)

14. Extra end posts for some shelf racks were not ordered. The shelf racks were connected with common end posts in the old plant, but were separated in the new plant. The shelf racks cannot be erected on moving day because the extra end posts are missing. As a result, the stock for the shelves cannot be loaded and is scattered all over the floor. (This error applies to pallet racks also.)

15. The dimensions of the equipment involved in a close fit between two walls were not double-checked. Neither was the distance between the walls. On moving day, the equipment does not fit, and some of it is moved into the aisle. Now the area is crowded and not well planned.

16. A consolidated power drop was not specified for the office bullpen area. Some power lines were randomly dropped from the ceiling, and others were laid on the floor. The floor lines are covered with plastic runners or taped to the carpet, creating a safety hazard and a constant inconvenience.

17. _____

_____

_____

_____

_____

_____

Figure 11.1   Errors checklist (*Continued*).

a minimum of piping required and a central location for all employees. The height of the station was specified to accommodate someone in a wheelchair.

Figure 11.3, the design tips checklist, should be used as a reminder of the many suggestions that can help you design a commendable layout. These tips differ from the previous two checklists in that they may not be required to complete an adequate layout project, but they will assist in creating one that is outstanding and represents a state-of-the-art product.

This ends the preparation phase of the entire project. The rest of the book will be devoted to the mechanics of designing a plant layout, office layout, and utility overlays, and planning the relocation on moving day.

The omissions listed below are features of the plant that are normally included in routine plant layouts, but may be forgotten during the hectic activity of the design phase. Review this list for items you may have overlooked, and add others that apply to your new plant.

1. *Lockers.* Provide lockers where direct labor employees can store their coats and other personal gear during working hours. "If locker rooms are not provided, OSHA requires facilities to be provided for hanging outer garments"[1] (p. 1797). Lockers are much better. Look for hallways at least 6 ft wide where you can mount the lockers along one wall.

2. *MRB storage area.* The Material Review Board (MRB) is responsible for disposing of parts and materials that have been rejected by production. This board meets regularly, determines why the parts were rejected, and decides what the short- and long-term solutions are. A storage area is needed for the rejected parts until their disposition is complete.

3. *Women's rest area.* This feature is in addition to the women's rest room. An OSHA requirement[1] (p. 1796): "Where ten or more women are employed, at least one resting area is required. Two beds are required if between 100 and 250 women are employed, and one additional bed is required for each additional 250 women." At the minimum, a rest area should have a comfortable chair.

4. *Lunchroom.* OSHA requires[1] (p. 1796) that "in all places of employment where employees are permitted to lunch on the premises, an adequate space suitable for this purpose be provided for the maximum number of employees who may use each space at one time. Such space shall be physically separate from any location where there is exposure to toxic chemicals."

5. *Floor stock rack.* In most plants, hardware or other common parts are stored in a rack near the production area. There is free access to these parts, so a work order is not required for their issue. This saves time and paperwork for the direct labor employees. This rack should be near the storeroom, since all this material is processed through the inspection and stores area before it is placed in floor stock.

6. *First aid and nurse requirements.* The first aid supplies and facilities required will depend on the exposure of the employees to industrial hazards. At a minimum, a complete first aid cabinet should be readily available to all employees. "For each 500 employees, an industrial nurse should be hired"[1] (p. 1796). If a nurse is required, assign to the nurse a room or area on the layout. See your first aid supplier for equipment recommendations.

7. *Maintenance shop.* Even a small plant needs a room where engineers and technicians can perform minor repair work. A workbench, large vise, drill press, and tool cabinet are the minimum equipment required. Larger plants that have a maintenance staff will require a complete facility capable of extensive construction projects.

8. *Fire extinguishers.* They should be located on the layout no more than 50 ft from any employee, and they should be clearly visible. No

Figure 11.2 Omissions checklist.

closed doors are allowed between employees and their assigned fire extinguishers. If an area is especially hazardous, the extinguishers should be located near it, but far enough away to be outside the fire zone. The fire extinguisher should be rated for the type of fire anticipated. See your supplier for specific situations in your plant. Mount them with the top 4 ft off the floor for easy access from a wheelchair. Make sure there are enough fire alarms in the building also.

9. *Drinking fountains.* They are normally included when a building is constructed and should probably be within 100 ft of every work station. The water jet should be 36 in above the floor, or mount a paper cup holder that can be easily reached by a worker with a disability.

10. *Eyewash stations.* If safety glasses are required in the production area, then eyewash stations should be included also. The stations should be located on the layout no more than 100 ft (10 seconds) from any employee in a low-hazard operation and 25 ft in a high-hazard operation. Gravity feed models are available that are not connected to the water supply. Plan on how a worker with a disability will use the station. See your supplier for more information about specific situations in your plant.

11. *Emergency lighting.* In case of a power failure, some sort of automatic battery-operated lighting system should be provided throughout the plant. Don't forget the stairways and elevators. See your supplier for specific situations in your plant.

12. *Pass-thru.* An opening with a counter top in the wall of a secured area is sometimes desirable if communication without access is needed. An example might be between the incoming inspection room and the storeroom. A door would not do, however, since neither group wants members of the other group in its area, even momentarily.

13. *Dirty floors.* Some operations are inherently messy, and the floor of the room is impossible to keep clean. If that is the case in your plant, a temporary floor covering may be the only practical solution. Used carpets are readily available, can be cut to size, create a padded work area, and are free. See your local carpet installer for a limitless supply. Indicate the carpet outline on your layout.

14. *Sprinkler system constraints.* Do not specify storage racks so high that routinely held cartons will interfere with the operation of the sprinkler system. Even if the sprinklers are located in the aisles, be sure the cartons do not interfere with the overall water-spray coverage of the storeroom.

15. *Hot water pipes.* If possible, have the hot water pipes insulated throughout the plant. It not only will save energy, but also will provide more immediate hot water wherever it is needed.

16. _____
_____
_____
_____
_____
_____

Figure 11.2 Omissions checklist (*Continued*).

The tips listed below will help you fine-tune your plant layout. Review them often during the layout design phase, and add more tips as they come to mind.

1. *Windows.* Always include the window locations on the new plant structure drawing. The window locations will usually affect the placement of equipment in the immediate vicinity. Avoid placing equipment that will interfere with the window openings. To avoid blocking the windows, locate aisles along the wall next to the windows.

2. *Washing fixtures.* If laundry tubs or sinks are needed in the production area, try to place them in the vicinity of other hot water lines. This not only will save money on the plumbing work, but also will provide hot water more quickly.

3. *Storeroom counter.* If an issue counter is required for the storeroom or tool crib, avoid using a divided door with a shelf in the lower section. It is much more convenient to have an opening in the wall with its own counter and a separate security door for access. This avoids the common inconvenience of having to issue parts while someone else is trying to enter or leave the room.

4. *Internal windows.* If you design a room within the confines of the new building, try to incorporate windows on the internal walls. This will allow the employees in the room to feel that they are part of the total group. It will also make the room easier to supervise and will allow visitors to observe the operation without entering the room.

5. *New wall location.* If it is convenient, try to place the new walls so they enclose the columns of the building. This presents a better appearance for the area, and the columns will not interfere with your layout. However, do not be overzealous and compromise a good layout by moving a wall over to the nearest line of columns. The function of the layout comes first; then examine the columns to improve the appearance.

6. *Electrical supply drops.* Try to specify that the power drops to a line of workbenches be located on a column. This will involve placing the conduit right on the column, which must have at least one bench next to it. If that is not possible, use one consolidated power drop for each group of benches. If you do not plan your drops, the electrician may install a power bus above a row of benches and drop a power cord to each bench. As a result, the production area will look like a forest of black power lines. This tip applies to office layouts also.

7. *Drop ceiling.* If your plant has a high ceiling with many unsightly ventilation ducts and other structures in the overhead, the appearance and working environment can be greatly improved by installing ceiling tiles. Consult with your maintenance group about providing access to the utilities above the drop ceiling. This installation will be the most prominent improvement that you can make in the new building.

Figure 11.3. Design tips checklist.

8. *Material handling.* Investigate the most current equipment available for handling your parts, units, and final product. Here are some samples of the many possibilities:

   a. Conveyors can have conditional sensors so they can stop, start, change speed, and route containers automatically.
   b. Automated storage/retrieval systems can be used to pick and deliver parts to an assembly operation instead of manually kitting the parts in containers. Have your parts containers designed for automatic kitting.
   c. Custom work holders can be designed to travel with the unit yet lock to the conveyor at each work station and assist in the assembly operation.
   d. Your 4Q vendors can deliver parts directly to production on rolling carts, or they can load parts feeders for you. These holders are on-location storerooms supplying parts directly to the assembly operation.
   e. Products can be redesigned to be transported and manufactured with more automated equipment.

9. _____
   _____
   _____
   _____
   _____

Figure 11.3.   Design tips checklist (*Continued*).

12

New Plant Layout

This chapter will describe how the IE brought together all the charts, analyses, diagrams, improvements, and drawings into the design of the initial new plant layout for the Flexible Division. It will also present some basic dimensions that you can use as guidelines to design your initial new plant layout. Instructions for designing your layout will also be given, using the Flexible Division layout as a guide. Concerning the plant layout project schedule in Fig. 6.1, task 4, *GERRY B.: Set up prototype assy. line* was completed just as the IE started to design the new plant layout.

In this discussion, many figures will be referred to, and it will be easier to follow the text if you make copies of these drawings:

1. Fig. 3.1, List of improvements
2. Fig. 4.2, New CEDAR Labor Analysis
3. Fig. 5.1, New plant equipment list
4. Fig. 8.1, Settings for layout drawing
5. Fig. 8.3, New plant drawing
6. Fig. 10.3, Old plant layout
7. Fig. 10.4, Old plant equipment numbers
8. Fig. 10.5, Old plant material flow
9. Fig. 12.1, New plant layout
10. Fig. 12.2, New plant equipment numbers
11. Fig. 12.3, New plant material flow

Old versus New Comparison

When the IE designs the new plant layout, it is the culmination of all the effort that preceded it. The flowchart, the labor analysis, the improvement list, the relationship diagram, the old plant layout, and the new plant drawing all play an integral part. You can now compare the various layouts made by the IE and immediately realize how the Flexible Division's new plant layout evolved from the earlier old plant layout and the new plant drawing. By examining this evolution closely, you can, in effect, assume the role of the IE and evaluate the choices that were made in designing the new plant layout.

New Plant Drawing and New Plant Layout

Figure 12.1, the new plant layout, is the final version, and it illustrates the Flexible Division new plant layout with the work centers labeled. Compare this figure with Fig. 8.3, the new plant drawing. Notice the structural differences between these drawings, and how the original building is still the predominant edifice. As you can see on the new plant layout, the wall configuration of the original new building was changed to correspond to the work center relationships of the Flexible Division.

These additions and deletions are typical of the modifications normally made during the remodeling of a new building. In this case, they enable the new flow pattern to function efficiently, and actually simplify what was before a complicated and unwieldy manufacturing system. Notice the operation numbers next to each assembly bench. They are circled and conform to the new sequence described in Fig. 4.2, the new CEDAR Labor Analysis.

In addition to the wall changes, a counter for issuing parts from the storeroom, many door openings, and conveyor passageways in the inspection area have been added. A fire escape door has been installed in the north wall, correcting a serious deficiency of the original new building.

Old Plant Layout and New Plant Layout

Compare the new plant layout (Fig. 12.1), with the old plant layout (Fig. 10.3). Although the old plant is not drawn to scale, the areas of each work center can still be roughly estimated. The total areas of both plants are about the same, but the allocation has changed considerably. The new storeroom is smaller, since critical parts are now supplied by 4Q vendors, but all other areas are the same or slightly larger. The assembly area of the new plant is about the same size as the old, but it is far more organized and capable of a much higher capacity.

By using a simple gravity flow conveyor, the old batch assembly process has been upgraded to a sequential assembly process. A break or conference room has been added to the new plant as well as a change room leading into the thick-film clean room.

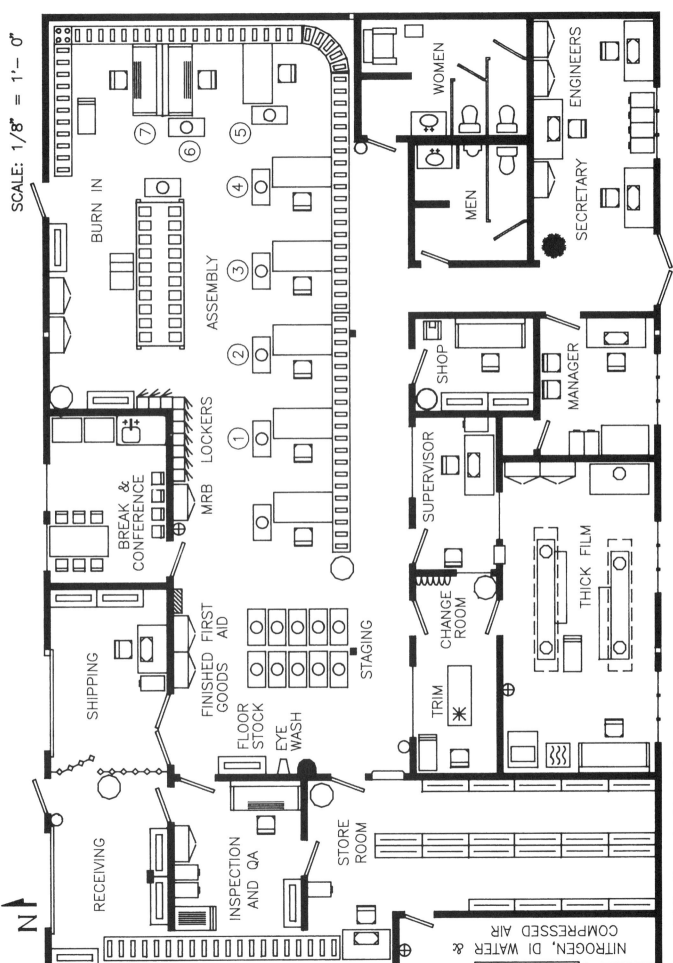

SCALE: 1/8" = 1'– 0"

Figure 12.1 New plant layout.

The new thick-film area will be a class 100 clean room with a much greater capacity. The new furnace and laser trimmer will increase the output, improve the product quality, and expand the range of products that can be created in this facility.

Old Plant Equipment Numbers and New Plant Equipment Numbers

Figure 12.2, showing new plant equipment numbers, illustrates the Flexible Division new plant layout with all the assigned equipment numbers. These numbers were assigned in Chapter 5 and are shown in Fig. 5.1, the new plant equipment list, and then again in Fig. 10.4, the old plant equipment numbers. Compare these three figures and notice how the entire complement of equipment, old and new, was initially entered on the list, recorded on the old plant layout, and finally located on the new plant layout.

Select, at random, certain pieces of equipment, and follow them through this sequence of events, observing how the IE planned each step and arrived at the final layout. This new plant layout is not the first attempt designed by the IE. Rather, it represents the results of intensive design work by the IE plus many one-on-one meetings with all users and a consolidation of all information. In the end, the design of the new layout satisfies the users' needs, incorporates the smooth flow that was envisioned by the IE, and introduces production innovations created by the IE. All these items are reflected in the new plant layout.

Old Plant Material Flow and New Plant Material Flow

Compare Fig. 10.5, the old plant material flow, with Fig. 12.3, the new plant material flow. The smoother flow of the new layout is obvious. Also, the reduced number of benches shows that the labor cost has been improved. After examining the flow pattern, you might expect the product cost to improve in the same proportion as the labor cost. It does not. It is true that the direct labor hours are sizably less for the new plan, but the labor cost of the memory bank is only 8 percent of the total product cost. The parts and other expenses required to build the memory bank account for 92 percent of the product cost. That's a typical cost ratio, and it must always be up front during any cost-saving project. For high-tech products, direct labor hours are relatively unimportant. Of course, your product may be just the opposite.

It is the overall flow time reduction that is of paramount importance for this plant layout project, the major reason for instigating the project in the first place. The flow time determines the reaction time of the system, the flexibility of the system, and the quantity of finished goods required to support the system. All these items contribute significantly to the future profit position of the company.

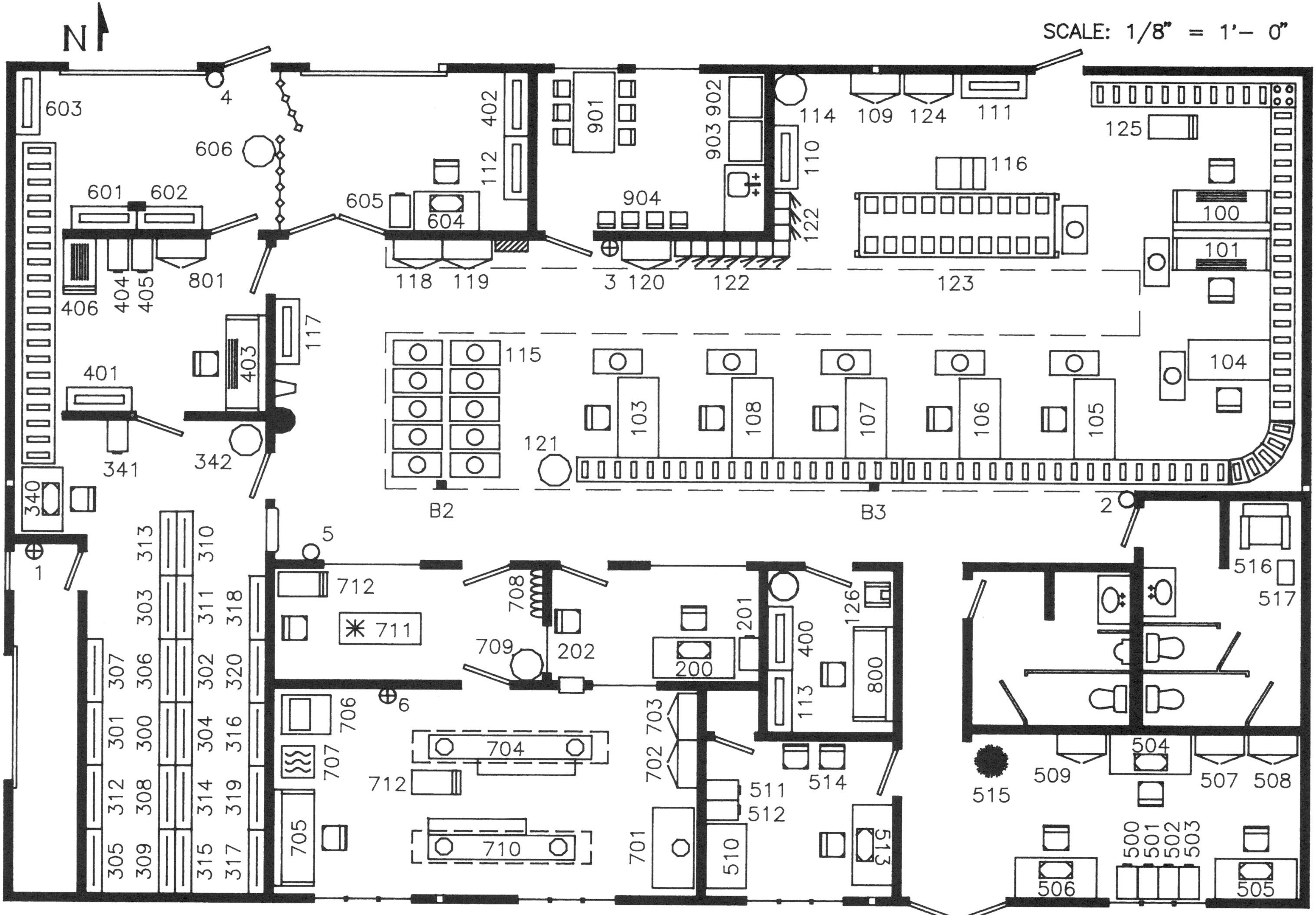

SCALE: 1/8" = 1'– 0"

Figure 12.2 New plant equipment numbers.

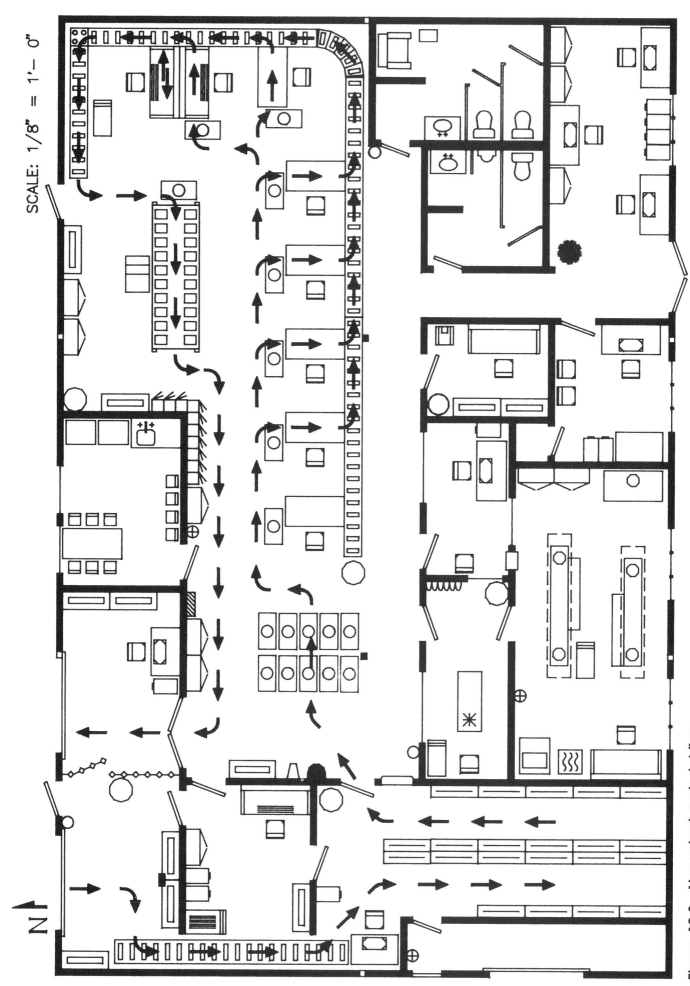

SCALE: 1/8" = 1'– 0"

Figure 12.3   New plant material flow.

Old Plant Layout and List of Improvements

The improvements that were incorporated in the new plant layout are described in Fig. 3.1. Compare this list with Fig. 10.3, the old plant layout, and Fig. 12.1, the new plant layout. Notice how the improvements were combined to create a new and more efficient manufacturing system. Bear in mind, this is not an example of a highly refined and automated assembly plant that has been developed after years of experience. Rather, it is a typical assembly operation that is common in the high-tech industry for a state-of-the-art product. Each plant layout will present a different set of circumstances and will require a different set of improvements. Your list of improvements will reflect the current state of advancement of the production process in your branch of the manufacturing industry.

The improvements and minor problems that the IE accounted for in the new layout are typical of the items that you will encounter also. If you are interested in following the IE through this process, here are a few of the items that can be easily recognized:

1. An equipment number series is assigned to each work center. Numbers 100 to 199 are assigned to production, 300 to 399 to the storeroom, and so forth. This plan is very helpful for identifying the equipment in the old plant while you are making the new plant layout and while you are reviewing the new plant layout with the users.

2. In the assembly area of the new plant, the benches are arranged in line and are connected by a gravity conveyor to provide a sequential assembly system. A dedicated staging area is provided so assembly carts can be loaded and moved into production on demand. This is especially desirable to provide the high degree of flexibility required for this line of products.

3. In the shipping and receiving area of the old plant, shelf units 601, 602, and 603 have common end posts. When 603 is separated from the section of shelving for the new layout, an additional end post is purchased for it. Shelf unit 402 is moved from the inspection area in the old plant to the shipping area in the new plant but still retains the original equipment number. In that way, both supervisors can readily understand and approve of the plan.

4. As explained earlier, a QA station is no longer needed in the new production plan, so bench 800 in the old plant is moved to the maintenance shop of the new plant. The QA employee who trains the operators to perform their own inspections is based in the incoming inspection area and uses cabinet 801 and new rolling cart 406 to train new operators and monitor product quality.

5. In the thick-film area, new laser trimmer 711, which does not require a clean room, is installed in the change room to provide easy access for both the clean-room personnel and the engineer who programs the trimmer.

6. In the thick-film production area, new conveyor furnace 710 is installed and dedicated to firing conductors. Old furnace 704, which had to be re-profiled to fire resistors after firing conductors, is now used to fire resistors only. The overhangs at each end of the furnaces are shown as dotted lines. They are removable for easier maneuverability. Furnace 710 is the new unit, and it will be moved into the room first.

7. In the women's rest room, easy chair 516 and table 517 were purchased, and enclosed for privacy, because more than 10 women are employed in the plant.

8. A combination break and conference room with vending machines and a sink is added to the new plant. In addition to being used for lunch and break periods, this room is used for qualified business meetings.

9. In the production area, workbench 103 is used for training and for future memory products that may require more assembly operations.

Dimensional Guidelines

You can use Fig. 12.4, the plant layout guidelines, as a guide for some basic dimensions of your initial plant layout drawing. These guidelines will help you get started on the design without getting bogged down with minor decisions. Later on, as the design is reviewed by the users, you can modify the dimensions to suit the conditions in your plant.

You can test the clearance dimensions for a particular situation in your plant by arranging some actual equipment on the floor and simulating how this equipment will be used. To test the aisle widths, measure off some aisles and duplicate the traffic conditions of the new plant. In any case, the guideline dimensions can be modified to conform to your plant requirements.

The numbers of rest room facilities are general guidelines only. The legal requirements may change in the future, or they may even vary in your local community. If your rest room facilities are definitely insufficient for the number of employees in your plant, consult with an architect before you make extensive renovations.

Initial Setup

At last, all the preliminary steps leading to the plant layout design have been concluded, and you are ready to start laying out your new plant. At this point, if you have followed the sequence of design events given in this book, you should have a firm idea of the function of each work area and where it will be located. You may also suspect that one of those areas will be more crowded or restricted in some way than any of the others.

You should lay out this critical area first. In that way, if the equipment will not fit the space in a smooth and organized man-

The guidelines suggested here should be evaluated individually for every application. They have been found to work in practice, however, and they can be used to start the design process. In general, these dimensions conform to the Americans with Disabilities Act.

1. Clearance required behind operator's side of workbench (add 1 ft for wheelchair access)
   a. Operator access only — 3 ft
   b. Operator and occasional passerby — 4 ft
   c. Operator and regular use by others — 5 ft

2. Aisle widths
   a. For occasional traffic in the production area — 3 ft
   b. For occasional traffic within an office area — 3 ft
   c. For light traffic in a perimeter aisle — 4 ft
   d. For light traffic around a bullpen office area — 4 ft
   e. For rolling carts in the production area — 4 ft
   f. For regular traffic in the production area — 5 ft
   g. For medium traffic between enclosed offices — 5 ft
   h. For medium traffic in a perimeter aisle — 5 ft
   i. For heavy traffic in a perimeter aisle — 6 ft
   j. For heavy traffic between enclosed offices — 6 ft

3. Door widths
   a. For small storage rooms — 32 in
   b. For normal production areas and rooms — 3 ft
   c. For outside doors — 3 ft
   d. For rooms with oversized equipment — 4 ft
   e. For rooms with large equipment (double 3 ft) — 6 ft

4. Storeroom clearance between facing shelves
   a. Occasional use with minimum floor space — 33 in
   b. Regular use with minimum clearance — 3 ft
   c. Regular use with comfortable clearance — 4 ft
   d. Heavy use by many people — 5 ft
   e. Regular use with a cart; cart width plus — 2 ft

5. Rest room facilities: There should be at least one toilet stall for the disabled in each rest room. Rest rooms should be within 200 ft of all employees. The interior shall not be visible when the door is open. A urinal can be substituted for a toilet in the men's room, but two-thirds of total must be toilets.

| Number of women | Women | | | | Men | |
or men in plant	Sink	Toilet	Rest area	Bed	Sink	Toilet
1 to 9	1	1			1	1
10 to 24	2	2	1		2	2
25 to 49	4	3	1		4	3
50 to 74	7	4	1		7	4
75 to 100	10	5	2	1	10	5
15 more	+1				+1	
30 more		+1				+1
100 more women			+2	+2		

Figure 12.4 Plant layout guidelines.

ner, you can alleviate the situation before making other commitments. You can move the walls, divide the work center function, or reassign the work center to a different area of the plant. You can even add a mezzanine to provide more floor space, but that is a last-resort solution.

You can use your version of Fig. 10.4, the old plant equipment numbers, as a vehicle to retrieve the old and new equipment into your new plant layout drawing. Assuming that your computer is ready to use the AutoCAD program, retrieve your old plant layout drawing with the equipment numbers shown. Turn off the layer that contains the walls. Only the old equipment along with the equipment numbers should be left on the drawing.

Set the SNAP distance to 12 in. Make a block of the equipment that you will use to lay out the critical area, and WBLOCK it to a temporary block file. If you have made blocks for the new equipment or have equipment from other areas that will be used in the critical area, include them in this block also. QUIT this file and retrieve your new plant drawing, which is similar to Fig. 8.3, the Flexible Division's new plant drawing. Your version of this drawing will be used to start your plant layout.

Pan to one side of the outline so there will be room to insert the block of equipment that you just saved. Set the SNAP distance to 12 in. Insert this block next to the layout drawing, and explode it. At this point you should have the new plant drawing with all the walls, columns, and other restrictions and, beside it, the equipment and numbers needed for the critical work center layout.

Figure 8.1, the settings for the layout drawing, lists the layer assignments for the blank drawing file LAYOUT.DWG. If you started your old and new plant layouts using this drawing file, then the layers and colors for both drawings will be the same. When you explode the block, the equipment and numbers will automatically revert to their previously assigned layers.

The equipment for each area, such as assembly or storeroom, should be assigned a separate layer and color. In that way, you can freeze all the other work area layers except the one you are designing, and decrease the time required to redraw and regenerate. Layers FILL and NUMBERS are mentioned in macros that were described in Chapter 9. If you want to use these macros and you change the layer names, you will have to edit the macros.

At this point, decide if you are going to retain the equipment numbers while you are making the initial layout or to erase them and add them back later. If you are familiar with the equipment, it will be easier to just erase all the numbers and reassign them later when the equipment is oriented properly. You can always refer to the old plant layout drawing to identify the equipment and numbers.

Set the SNAP distance back to 3 in, and generally leave it there for all future layout work. Before you start placing equipment in the layout, it is best to accomplish some initial steps:

1. *Dimensional survey.* Examine your new plant layout drawing closely for overall dimensions. Are all structures of the

new building located correctly on your layout? This will be your last check before the design begins, and dimensional errors may be costly to correct after the next step is taken. Double-check the overall dimensions of the building and the column centers against your notebook data if you have any reservations about their accuracy.

2. *Aisles.* Start by deciding approximately where the aisles should be placed. Do not leave a large work center area closed without an aisle through it. Aisles are usually located along walls that have doors which lead to offices. Also, aisles can be placed beside the columns, and thus alleviate the problem of having columns within the work area. If there is a main door leading to a large area, then it is common to have an aisle along the wall on each side of the door and an aisle leading out onto the area in front of the door.

If there are windows on an outside wall, it is common to have an aisle there also. However, if you have a wall without windows and there are large pieces of equipment in the area, the equipment should be placed along the wall, with the aisles on the inside.

The aisles should be delineated by dashed lines on your drawing on the NUMBERS layer. See Fig. 12.2, the new plant equipment numbers, for an aisle-marking example around and through the production area. Terminate the aisle line 6 in short of any walls or obstacles for a better appearance. The aisle lines that you draw now will be used in the new plant, just before the move, to position the aisle-marking tape on the plant floor. Put the dashed lines only where you want the tape to be on the floor.

3. *Production cabinets and shelving.* In some cases, shelving will be needed in the production area to store tools or other items that are used regularly. Floor stock shelving may also be needed. Any other requests for production shelving should be discussed in detail with the production supervisor. If shelving is installed without having a definite assignment, it will only serve to collect production debris. All items that should be processed at the time, but could be postponed, will be placed on the shelf. It will fill to overflowing with this sort of material, which will remain forever as an eyesore and dust collector.

When shelving is required, it can be placed against a wall next to an aisle. Notice the floor stock shelves in Fig. 12.1 as an example. This arrangement uses the aisle to provide access to the shelves without loss of floor space.

Trial Plant Layout

If you have some doubt about how the equipment should be arranged in an area, you can make a trial layout in the open space outside the plant walls. This would also be a good chance to try some of the macros described in Chapter 9. Start with the most important piece of equipment, or the one used first in the process. Move the equipment to an open space so there is room to place the other associated equipment near it. Then move the next piece of equipment into position, in line with the first.

Continue the trial layout by arranging all the main equipment for the work center in a logical, straight-line sequence. Add the subassembly feeder lines also, but use all the room you need. Other criteria may drive your layout besides in-line assembly. For example, if the critical area is used for testing, arrange the work stations in the proper relationship to perform the tests. By not limiting the layout to the confines of the walls, you have more freedom to concentrate on function rather than location.

Do not worry about the spacing between the benches or the auxiliary items that may go with the benches. The object of this initial layout is to get started on the design phase and to solidify your concept of the function of the work center. By concentrating only on the main pieces of equipment, the logical location of each piece will become more obvious.

When you have the equipment arranged in the proper sequence, copy each piece into the actual room or area where it will be located. Now you can arrange the equipment to fit the confines of the room, while retaining the original sequence. After the main equipment and benches are in place, add the auxiliary items that will complete the area layout.

Ensure that every item from the old layout, plus the new equipment, is included in the new area layout. That was a primary reason for making an old plant layout. The old plant layout is a visual inventory of new plant equipment that can be easily tabulated and verified.

If the equipment does not fit easily into the new layout floor space the first time, your drawing may become hopelessly disorganized. Rather than continue to work within the real layout floor space, erase the equipment in the critical area and start over from your trial layout.

The floor area required for each work center can be calculated during the initial planning phase, but the final test for adequate room size can be made only after all the equipment has been laid out. If the area appears crowded, now is the time to alleviate that condition. On the other hand, if the area is too large for the equipment, you can reconfigure it for more efficient use. Other related functions could be moved into the area, or a different work center could be located there to better utilize the space.

A layout area will appear crowded when almost every square foot of the work center is designated for some function. Be sure to allow some space for the movement of people and material. Refer to the production area shown in Fig. 12.1, and notice how each bench has a minimal working space for the operator, yet the total area still has a somewhat spacious appearance. There are other ways to create this same feeling of spaciousness and yet use the space efficiently:

1. Make the aisles slightly wider.

2. Arrange the benches, machinery, or shelving so that the users are oriented back to back, sharing the open space behind them.

3. Provide enough space around all machinery for easy access.

4. Eliminate walls or move some equipment out of the area.

The object of this initial design phase is to arrive at an ideal plan that will satisfy every goal, without regard to cost or additional construction. This will free your mind from having to consider all the constraints imposed by the building or any political issues. Concentrate on and develop one aspect of the design at a time. Expand your thoughts by creating ridiculous solutions to your problems. Then extract those parts that have a shred of value, and use them to come up with still more ideas. This is a common method for developing original solutions. It is the basis for many synergistic meetings that are held for the sole purpose of developing new products. Always remember, plant layout is an art form; consequently, there may be many functional solutions to any layout problem. The final plan, however, will be a subjective decision made by you, with the help of the project team.

You will finally arrive at a layout that not only fits the space, but also has sufficient working room to perform the function of the work center. You may have to make compromises among the economics, politics, and structure of your plant in order to arrive at the most reasonable solution, but that is always the case. At this point, do not attempt to fine-tune the layout. Let it stand for now. You have solved the major problems, and that is all that is required at this stage.

Later, when you discuss your solutions with the users, more ideas and revisions will naturally evolve. It is always easier to change a layout than to design one from scratch. At this time, you should number all the equipment that you have placed in the layout. You can add a few flow arrows to your drawing, but save the detailed material flow plan until later, when the layout is more firm.

You can lay out the remaining work center areas in the same way as you did the first one. Refer to all checklists as each area is completed. The purpose of the initial layout is to start the design process, ensure that the equipment will fit the space reasonably well, and see that each work center will perform its function. Do not strive for a finished layout design at this juncture. Your time will be better spent on this endeavor later, after the user review process.

13

Office Layout and Final Approval

In this chapter, planning your office layout will be discussed along with other items that should be considered before you start the office design. The layout review with the users and the revision procedure will be explained for both the office and manufacturing areas. Finally, the approval and sign-off phase of the layouts will be described. Concerning Fig. 6.1, the plant layout project schedule, this chapter will complete task 9, *IE: Design new plant layout.*

Office Layout Considerations

The purpose of designing an office layout is threefold:

1. To organize the functional groups and equipment so their business can be conducted efficiently.

2. To ensure that the required equipment will fit into the allotted space with sufficient room to provide for the comfort of the office workers.

3. To implement improvements, mainly in information flow time.

Office layouts are similar to production layouts in that the space assigned must be analyzed for efficient use. Just as in a production layout, the motivation is flow improvement. In this case, information flow, not material flow, is the lifeblood of the organization. If this flow is smooth and a program to reduce the flow time is in effect, a minimum office staff will be required. Conversely, more office workers generally mean more paperwork and longer flow times. There are exceptions, of course, but usually personnel requisitions can be replaced by procedural streamlining.

Company policy will vary concerning how much office space should be assigned to each employee. One company may assign private offices to everyone. Another will have private offices for the supervisors and everyone senior to that, and there are many other variations in between.

If the offices are to be moved and a layout is required, discuss the office area allocation with the department heads. That is, how much floor area should be assigned to each person. This will solidify the company policy regarding the area of offices for each level of management. If necessary, write down exactly how many square feet each level or each person should have for an office area. Make sure that each department has the same area for equal-level employees. Some people relate the size of their office to their status in the company, and they take it very seriously.

If you assign office areas yourself in a random manner, the results can be disastrous. Serious disagreements can erupt, and you will be at the center of the controversy. If walls have been built, it may be too late to correct the problem, and the ill feeling toward you may be permanent.

Closets and private bathrooms can be another point of contention, since both are highly prized by high-level administrators. Notice the closet in the manager's office in Fig. 12.1, the new plant layout. It is inconspicuous but allows the manager to put away all those items that normally clutter up many high-level manufacturing offices. If these features are included in your office layout, be sure that they are allocated according to company policy.

Almost all large companies will have a bullpen area for some workers, with low permanent partitions or temporary partitions that provide at least the feeling of a private office. The permanent partitions will be about 4 ft high while the temporary ones may be 4, 5, or 6 ft high. The temporary partitions can be rearranged easily, and thus make the space more open or consolidated, depending upon the short-term needs of the company.

Once again, it is important to work with the users if you want your office layout to be successful. If you do each detailed office layout without consultation, the office workers will rearrange the furniture to suit themselves anyway, even though their preferred arrangement may not make the best use of the space. Much time and effort will be wasted, and you will lose the opportunity to effectively design people-oriented work centers. On the other hand, if you review the layout extensively with the users, including the use of 3-D views, there will be almost no adjustments made after the move.

Regarding the equipment for high-level managers, there is a unique type of desk that deserves honorable mention. It has the appearance of a normal desk but has one feature that, at first, does not seem worth the extra expense. Later, after becoming accustomed to using it, most managers will not part with it for any price.

The unusual feature of this desk is a shelf that slides out from under the desk top toward visitors. Its length is almost the full width of the desk, and it slides out about 2 ft. The result is an instant, large, clean working surface which visitors can use for drawings or paperwork and write on them freely. Best of all, everyone, including the manager, can see the table easily without getting up. If some managers in your plant are buying new equipment, this desk is worthy of consideration, even if it has to be custom-made.

The possible ill effects of radiation from computer monitors have been mentioned before, and should be a consideration in the design of your office layout. Most of the radiation emanates from the sides and back of the monitor. For that reason, do not design your office layout with anyone's workplace within 4 ft of the back or side of another operator's computer.

Older computers may require even more space than that. Notice in Fig. 13.1, a sample office layout, that no office worker is in the immediate vicinity of someone else's computer. If you must plan a layout for office workers in a minimum floor space,

SCALE: 1/8" = 1'- 0"

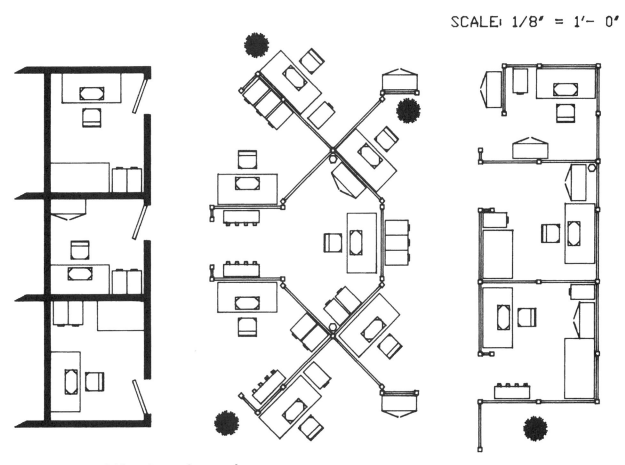

Figure 13.1 Office layout sample.

then obtain an electromagnetic radiation monitor, described earlier in this book. Measure the radiation, and ensure that every employee is working in a safe location when all the equipment is turned on.

Office Layout Planning

If you have little experience in office layout planning but want to expand your knowledge, tour the office areas of some progressive companies in your community. The time will be well spent. While you are there, notice and even make notes of the reception and secretarial area arrangements. If possible, ask the office workers what they like and dislike about their arrangements. In a short time you can become knowledgeable in the field of office layout and will be more prepared to create your own design.

You may have to construct an office relationship diagram if the information flow connections are not obvious. The best source of information for the diagram is the office workers themselves. Only they know the true interdependencies of each work center. For example, it is common for several people to share the same equipment constantly, such as the fax machine or the laser printer. Some may share their job responsibilities during breaks or other absences. There may be a great deal of traffic between two office areas that is not immediately obvious from observation.

If you take the time to become acquainted with a few key people and listen sympathetically to their needs, the benefits can be immense. The office layout can provide a very significant improvement in convenience if these interdependencies are incorporated into the design. If the office relationships are complicated, arrange them on a chart similar to Fig. 5.3, the relationship diagram.

Normally, the office areas will be part of an existing structure, and there will be some reluctance by management to relocate walls just for the office workers. Therefore, you should attempt to fit all equipment within the given confines of an existing office complex. Consider installing a pass-thru if the walls get in your way.

If you are going to make the detailed layout of each office, follow the same steps that you used for the production area. You may have to make an equipment list and a layout of the old, existing office, especially if there are many nonstandard pieces of equipment to be moved. In some cases, a list of the major equipment may be all that is needed. Regardless of which method you use to inventory the equipment, you must include the names of the people who use it. This is especially true of equipment that has multiple users.

Your list should show the item of equipment, its size, any nonstandard utilities it may require, and the people who will use it. There are many ways to prepare this information, and your company's circumstances will determine what is needed.

Before moving, management may direct all office workers to lay out their own new offices. In that case, you may still have to make an overall layout, but you can copy the detailed office layouts from the floor plan of each worker. It is possible that you may not have to lay out the details at all unless a record of the equipment location is required.

As you can see, there is a great deal of latitude in the layout of office spaces, but the key words here are *planning* and *cooperation*. Without either, it will be difficult for the organization to achieve its full potential.

Office Layout

It is beyond the scope of this book to describe all the various office, secretarial, and reception area arrangements that are possible, or other conditions that may affect the office environment. If you have a sizable office complex to lay out or if it is essential to specify the environment in detail, more information can be found in *Planning and Designing the Office Environment*.[17] This book not only covers office design and flow planning but also provides technical assistance for specifying acoustics, lighting, HVAC, and fire safety.

Once the occupants' names have been assigned to the offices, the detailed office layout can be started. As in the production layout, you can start with the aisles. Suggested aisle widths for office areas are shown in Fig. 12.4, the plant layout guidelines. You will have to estimate the traffic patterns and the traffic density for your plant. Here again, first locate the aisles on your layout, then consult with the users. They will know where high-traffic aisles are required and how wide they should be.

The Flexible Division office shown in Fig. 12.1, the new plant layout, was designed by the IE. Notice that the production supervisor, the engineers, and the secretary/receptionist are all located in close proximity. More individual office layouts are shown in Fig. 13.1, a sample office layout. A 3-D view of the same area is shown in Fig. 13.2. In the 3-D view, notice how much easier it is to understand how the office will function.

Several types of walls can be used to delineate offices in a bullpen. Bullpens with temporary partitions have the advantage of being readily adaptable to the needs of the office system. But bear in mind that temporary partitions can cost more than permanent walls with doors, and should be used only if flexibility is desired. Acoustical partitions or partitions with windows can cost even more. You should avoid the inexpensive type of temporary partitions with supporting legs that protrude out into the aisle or office space, because they can cause accidents.

If you use the conventional freestanding type of partition without legs, you will have to provide an anchor at the end of each long unsupported partition. This anchor should be a short partition at right angles, to provide stability. Notice the short anchor section at the end of some partitions in Fig. 13.1.

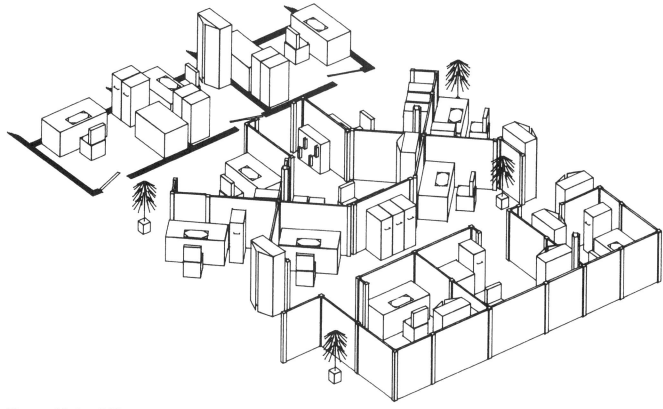

Figure 13.2   Office area, 3-D view.

When the bullpen arrangement is fairly permanent, the least expensive solution is a 4- or 5-ft-high permanent wall with unframed doorways.

The bullpen layout can be arranged in much the same manner as a series of enclosed offices. In Fig. 13.1, a conventional rectangular arrangement with partitions instead of walls is shown in the right-hand section. A free-form example of a bullpen with partitions is shown in the center section. An example of a conventional office with walls and doors is shown in the left section.

Your partition supplier can show you many styles of free-form office arrangements. If you use one with odd corner angles, be sure to buy a partition that can adapt to any angle at those posts where they are needed. Your supplier can also provide other helpful tips on how to use partitions successfully. All partition blocks in the equipment library are 5 ft high.

Once the departmental areas have been assigned, check your list of people who will occupy a given area, and enter their names on each workplace. Then check the square-foot area of each office to be sure that it agrees with the size requirements that you received from the department heads. If someone's office is too small or too large, this is the time to assess the situation.

Discuss the odd-sized floor space with the affected individuals directly. In most cases, it will be of little importance, or they will offer some solutions. To expand a crowded condition, you can place several people in one large office. In that way, their combined floor space will result in a more spacious area for each person. In any event, do not proceed until you are sure that everyone concerned with any irregular situation is satisfied.

The utility required for office equipment is generally a 110-V quad outlet. The computer cable outlets and telephone outlets should be placed near each power outlet. On permanent walls, the power outlets should be located about 8 ft apart.

On more expensive temporary partitions, a power outlet strip is built right into the lower portion of each unit, and the power drop that extends from the ceiling will be built right into the post. In less expensive partitions, the power outlets may have to be screwed onto the unit or simply placed on the floor. In these cases, the power, telephone, and computer line drop should be enclosed in a post that extends from the ceiling down to the floor. This can be painted to match the partitions. Notice the two power drops in the internal corners of the center view in Fig. 13.1. This power drop is included in the library of equipment.

The electrical contractor will want to know where you want the computer cables and power drops to the bullpens. The telephone contractor will also need this information so the wiring in the ceiling can be prepared beforehand. The amperage requirements will be needed so the electrical wiring can be installed in the ceiling before the move. Be sure to check any large equipment or portable heaters for a high-power 220-V requirement. Also check any 110-V heaters for a high-power requirement.

Notify the contractor about any nonstandard power requirements. If some equipment requires 220 V, notice the type number of the plug attached to the power cord. There are many types of 220-V plugs, and the contractor may have to order a special receptacle to match the type of plug on the equipment. If the special receptacle cannot be received in time for moving day, you can remove the receptacle from the old plant wall and install it in the new plant wall.

The actual installation of the power drop will be done after the partitions have been erected. Under no circumstances should you use, even temporarily, an electric wire on the floor across an aisle for the power source to a bullpen. It may be taped down or covered with a tapered riser, but it will still be unsightly and will create a serious safety hazard. Someone may trip over the cord, which could result in a serious accident. The layout designer may be held responsible, and the company may be liable for damages. If a temporary installation is unavoidable, route the wire overhead.

Plant Layout Review Preparation

It is assumed at this point that your initial plant layout and office layout are complete. Although you have invested a substantial amount of time and deliberation on the initial design, both these layouts should still be considered preliminary. That should not affect, however, the degree of formal preparation that you make to present the layouts to the users.

That is, the layout should be prepared in a ⅛ in = 1 ft scale, color-coded by work center areas with the columns and walls filled. It should look like the Flexible Division layout in Fig. 12.2, showing new plant equipment numbers. In the actual preliminary layout, the IE assigned colors to the assembly area, storeroom, and other areas to delineate them more precisely, and to better define the flow of material between work centers.

Figure 13.3, the layout legend, shows 2-D versions of all the equipment categories in the library. It can be used for the legend on your drawing. The block on the diskette is named LEGEND.DWG, and can be inserted, exploded, and edited just like any other block. The base point is in the upper left-hand corner. After inserting and exploding the block, you can edit it to match your drawing by removing some equipment blocks and adding others.

On most drawings, the legend will conform better to the drawing format if you arrange it in one long column down the side or one long row across the bottom of your drawing. The legend has not been inserted into the LAYOUT.DWG file.

Along with the new plant layout, you should also prepare your old plant layout for the review process. Your old plant layout should be similar to the Flexible Division layout in Fig. 10.4, showing old plant equipment numbers. You should insert the legend into this drawing also and edit it so that all equipment categories are represented. Ideally, you should plot your old plant layout with the same color coding as your new plant layout, so the observers can orient themselves to their work centers and identify their equipment more quickly.

Plot both layouts on opaque paper with colored pens, if possible. If you use felt-tip pens and an opaque paper with a slight sheen to the surface, the lines will not bleed and the colors will be more brilliant. Experiment with various pen widths during these early plots to arrive at the best width for each type of line and each type of drawing. Then, when you make the final plot for management, the layout appearance will be perfect.

Choose your pen widths carefully, as they can make a significant difference in the presentation quality of your layout, especially for 3-D drawings. Most felt-tip pens come in two widths but can be modified to make four widths quite easily. You can widen the tip of any pen by hand before using it. Simply bear down and draw the tip across a piece of paper until it spreads out to the desired size.

Before you start the review process, generate the 3-D view of the work center. As mentioned earlier, the viewpoint of 10,-8,9 will result in a view that is optimized for most layouts. To start your 3-D view, enter the VPOINT command and answer with the numbers 10,-8,9. A 3-D wire-frame view of the complete drawing will be generated. You can then zoom and pan until the exact view you want is on your screen. Be sure to name the view so you can restore it easily and plot it later.

Notice that only the assembly area is shown in Fig. 13.4 and that all other area layers have been frozen. The 3-D view of the

Figure 13.3   Layout legend.

entire new plant is shown in Fig. 13.5. The IE showed the over-all view of the plant to each user, as well as the enlarged view of each person's work area, so that more details could be shown when people reviewed the layout. You can, of course, plot more of the plant on larger paper to accomplish the same end, but A-size plots of 3-D views are more convenient for

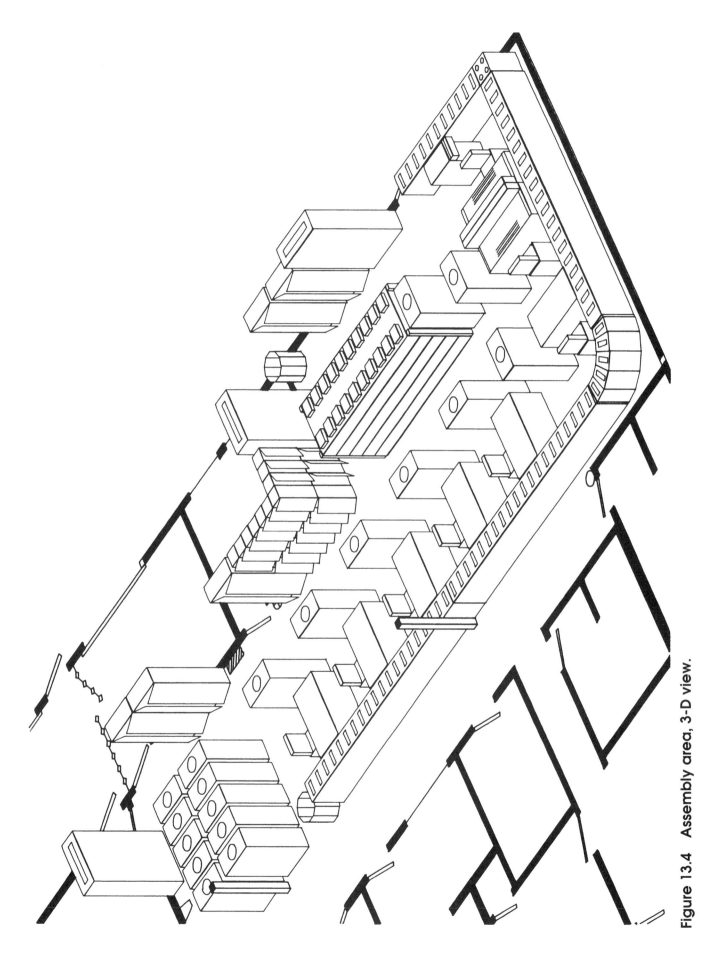

Figure 13.4 Assembly area, 3-D view.

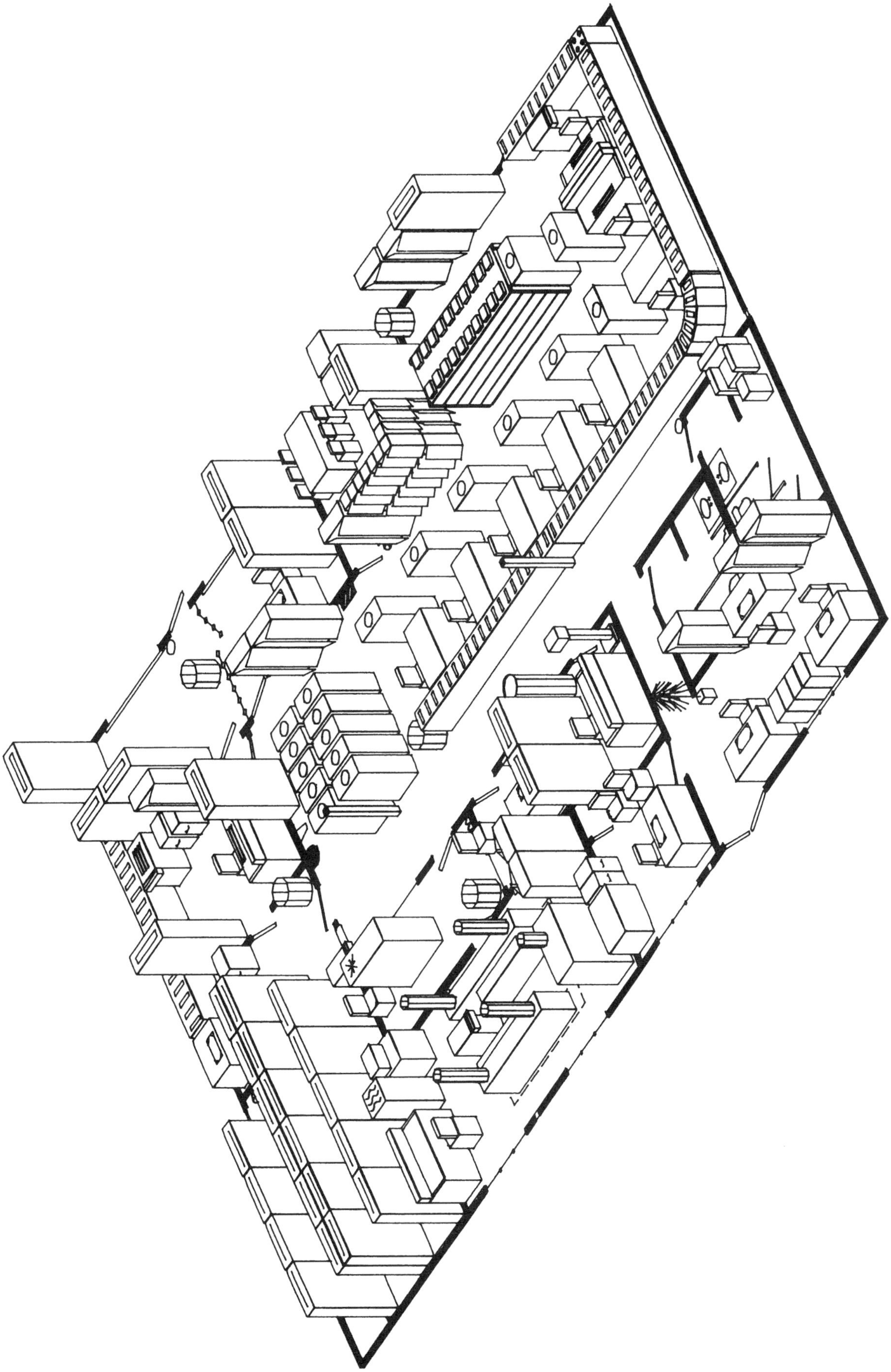

Figure 13.5   New plant, 3-D view.

everyone to pass around and file. Also, the user is usually concerned with only one work center area at a time during the review process.

After you have frozen the unneeded layers, zoom in on smaller sections of the same view and make sure that each block is displayed in 3-D, especially those blocks that you have constructed yourself. If you find one that is "flat," zoom in on it and change the thickness while you are viewing it in 3-D. If the block is used elsewhere in the drawing, make a new block of it.

When all the equipment in the 3-D view appears to be displayed correctly, use the HIDE command to change the wire-frame display to a true 3-D display. The speed of your computer processor and the size of your drawing will determine how long it takes to process the image. It can vary from seconds to nearly an hour. When the finished image is displayed on your screen, again check all the blocks for accuracy. Do not zoom in on this display, or it will revert to a wire-frame view.

Plot your 3-D view, using the same pen colors that you used for the plan view. To plot your 3-D view, plot the view, hide the lines, and plot to fit. Here again, the time to plot the drawing will vary greatly, depending on the speed of your system.

If your computer runs out of memory during the 3-D plot, it may exit the AutoCAD program back to DOS without warning. If that happens, start the program over, load your drawing, and review it for extruded circles. Since extruded circles take more memory in 3-D, replace them with extruded hexagons or octagons. Always keep your drawing as simple as possible.

Before you plot again, exit to DOS, start AutoCAD, and plot the 3-D view immediately, without viewing the file. That will allow the program to use more of its memory for the plot. If you still do not have enough memory to support the plot, break up the 3-D view into smaller areas.

Review, Revision, and Ratification

Review

Before starting the plant layout review process, you should have five documents to show the users:

1. The old plant layout with equipment numbers   (Fig. 10.4)

2. The new plant layout with equipment numbers   (Fig. 12.2)

3. The appropriate area layout, 3-D view   (Fig. 13.4)

4. The entire plant, 3-D view   (Fig. 13.5)

5. The equipment list   (Fig. 5.1)

Each work center layout should be reviewed with the users, not only the supervisors and other key people but also many of the workers. The depth of this review process is at the designer's discretion. If you find that most of the employees have many valuable ideas, then expand the approach to include even more people and more ideas. A major goal of the layout is to achieve maximum efficiency of material flow, and the layout review process is a means to achieve that goal.

On the other hand, and this is the most common reception, if everyone appears to accept the new layout and contributes only minor suggestions, you may think you can proceed without further investigation. Do not be tempted. Many people will not concentrate sufficiently on the function of their work center or work station and thus will not readily contribute to the review process. You should encourage a response, good or bad, from every person you interview.

Don't make the interview just a question-and-answer session; start a discussion. Begin by asking direct questions about the flow of material through their areas of responsibility. Ask both workers and supervisors about certain details of your layout, for instance:

1. Should we kit the parts in the storeroom or deliver them in bulk to each station?

2. Should we test the units before burn-in or turn them on in the rack instead?

3. Should we assemble the units ourselves or have an outside vendor do part of it?

4. Where should we dispose of the packing material, in the storeroom or on the production floor?

5. If we unpack parts in the storeroom, how do we handle sensitive printed-circuit boards?

Assume that all your premises for designing the initial layout may be wrong. Ask ridiculous questions if the interviewee is still reluctant to participate. At the least, extract a few positive comments about the flow of material on the new layout. Do not let anyone be turned off by the plan view. Show the 3-D view from the start, ask questions, and point to the area of concern on the drawing. Good ideas are in the mind of every user, and it is your job to get the users involved and to expose their ideas.

Revision

If changes are to be incorporated in the new layout, make them with all the layers turned on, at least to start with. Then freeze those layers that will not be affected by the change. As you move equipment to a new location, be sure to move the equipment number with it.

When each area is complete, turn on the layer that contains the title of that area, to make sure that it can still be seen clearly. If you add new equipment, add it to the legend also.

After you revise the layout, review it again with the key users. Only minor changes should be required after this second review, but some complicated layouts may require several revision cycles before they are acceptable to everyone.

Ratification

The ratification process consists of a review by the department managers and, if necessary, a formal sign-off. You and each area supervisor should meet with the appropriate manager to explain the advantages of the new layout compared to the old. You should not have to "sell" the layout to the manager or anyone else. If you feel that anyone, even the president, may want to get involved in the layout design, offer that opportunity well ahead of time. This review and sign-off should be only a formality.

You can use all your drawings to explain the function of your new layout to the managers. The Flexible Division IE used these documents:

1. The list of improvements (Fig. 3.1)
2. The old plant layout (Fig. 10.3)
3. The old plant material flow (Fig. 10.5)
4. The new plant layout (Fig. 12.1)
5. The new plant material flow (Fig. 12.3)
6. The appropriate work center, 3-D view (Fig. 13.4)
7. The new plant layout, 3-D view (Fig. 13.5)

Since the supervisors have been involved directly with the layout development, most managers will defer to their judgment. The discussion at the brief meeting will center on satisfying the manager that everyone in the department, including the workers, approves of the layout.

This is the final phase of the equipment placement in the plant layout. After the managers' sign-off, there should be only minor changes requiring no further approval.

14

New Plant Alterations

In this chapter, directions will be given for altering the new plant to accommodate the new production plan and conform with the new plant layout. Additional drawings will indicate what revisions will be needed to modify the walls, add new electrical services, and install new mechanical equipment. This chapter will complete task 10, *IE: Specify new walls and utilities*, and task 11, *CONTRACTOR: Renovate new building*, in Fig. 6.1, the plant layout project schedule.

Alteration Preliminaries

Now that the new plant layout has been approved, you can proceed with preparing the plant for the new production system. In the case of the Flexible Division, the walls of the building must be modified to agree with the plant layout, and other changes will also be needed. These include new electrical services, new mechanical installations, and a new heating, ventilating, and air conditioning (HVAC) system for the microcircuit area. All these changes and additions fall within the category of plant alterations and are usually performed by an outside general contractor.

The Americans with Disabilities Act (ADA) of 1990 will influence the construction or remodeling work of your plant. There are five titles to the act, and each has specific regulations:

Title I: Employers with 15 or more employees—regulations for employing people with disabilities.

Title II: State and local government programs—regulations for government buildings and public transportation.

Title III: Public accommodations—regulations for buildings open to the public and buildings used for private enterprise.

Title IV: Telecommunication companies—regulations for telecommunication systems to accommodate individuals with hearing and speech impairments.

Title V: Miscellaneous legal provisions—regulations making it illegal to retaliate against individuals who exercise their rights under the ADA.

Nongovernment manufacturing and warehouse facilities fall under Title III. Therefore, new construction and plant alterations must comply with the regulations specified under the ADA Accessibility Guidelines unless it is structurally impractical. In addition, companies with existing facilities should comply with the ADA to the extent that is required to accommodate employees with disabilities. Compliance for these employees is not mandatory, however. It is required only if reasonable accommodation needed by the employee does not create an "undue expense" for the employer.

Be aware that expenses incurred for the required removal of physical or communication barriers for these employees can qualify for a tax write-off. Small companies can receive a direct tax credit for about half their expenditures for required alterations, up to a maximum of $10,000 per year. Large companies can receive a deduction for the entire expenditure, if it is required, in the year that it was incurred. New construction expenses are not eligible for these tax benefits.

Regarding your plant layout, all new construction and all remodeling of existing buildings must comply with the regulations. If you are working with an architect, then the architect should have designed the new building or alterations with ADA compliance in mind. However, discuss the construction with the architect, and make sure that the guidelines have been followed in the building. If they have not, consult with management about it. Also, keep the architect fully informed of any modifications or additions that you plan.

If you are redesigning an existing plant, as the Flexible Division is, then you and the contractor will have to ensure compliance. Most contractors are fully aware of the regulations, and you should be, too. If you want to learn more about the ADA requirements, several publications are available free, or at a low cost:

1. *Nondiscrimination on the Basis of Disability by Public Accommodations and in Commercial Facilities; Final Rule,*[12] ADA Accessibility Guidelines (ADAAG), about 140 pages. This is a complete guide to the ADA covering background, definitions, legal aspects, responsibilities, implementation, guidelines, and dimensions.

2. *Americans with Disabilities Act Accessibility Guidelines Checklist for Buildings and Facilities,*[16] ADAAG Checklist, about 180 pages. This is a detailed breakdown in checklist form of the guidelines section of the reference in item 1.

3. *The Americans with Disabilities Act Checklist for Readily Achievable Barrier Removal*,[19] 14 pages. This contains a brief list of "readily achievable" barrier removal solutions. It is an excellent guide to upgrading an existing building without incurring "undue hardship." Problems and possible solutions are given.

The Flexible Division decided to conduct a barrier removal program in conjunction with the new plant alteration work. The IE ordered item 1 from the preceding list in order to be more fully informed about the ADA in general. Since the modifications to the new plant were routine, the IE ordered item 3 as a guide for the alterations and the barrier removal program. With the checklist in hand, the IE toured the new plant and checked off yes or no to every item on the list. From that time on, the list formed an organized basis for short- and long-term modifications that would be required for readily achievable barrier removal.

If your building will be extensively altered as part of the plant layout project and an architect will not be retained, your responsibilities may be far more comprehensive. Expert advice should be retained, as mistakes can be costly and can even lead to some form of litigation. An excellent place to start and a source of information is your Regional Disability and Business Technical Assistance Center.[20] These 10 federally financed organizations have centers located throughout the country. Your regional center can supply you with all the referenced ADA material, the tax benefit information mentioned earlier, and more.

One of the best sources of information for removing barriers in your plant is your own employees with disabilities. If you have any, consult with them extensively about what should be done to eliminate restrictions on their freedom to operate in your plant. Terry, the Flexible Division's final assembly technician, uses a wheelchair and made many useful suggestions during the layout review phase. One was a request that the fire extinguishers be mounted on the wall with the top just 4 ft off the floor. In the old building, Terry could not lift the unit off the bracket because it was too high.

When the employees in your plant make requests of this nature, be sure to keep accurate records of the discussions. If an employee with a disability asks for specific modifications in the plant or to a work station, have the employee submit the request in writing and sign it. Before the modification is started, write down what work will be performed and have the employee approve the plan. Litigation involving ADA noncompliance is often the result of a simple misunderstanding between the employee and the company. Having everything written down and signed by the parties involved will go a long way toward eliminating that misunderstanding.

During the discussion with management about what will and will not be done for the barrier removal or alteration project, some modifications will be rejected because of cost. Keep records of these discussions also, and the basis for these cost decisions. Review the records with the parties concerned to be sure that everyone agrees with the basis for the rejection. Later on, a new employee may challenge these very decisions, and written records will be invaluable if litigation should result from that episode.

Wall Modifications

The first step the IE took to alter the new Flexible Division plant was to create the drawing in Fig. 14.1, showing wall modifications. This drawing indicates what structural changes will be needed in the new plant. It is also an example of what you might design as a guide to the general contractor for your new plant modification. Remember, if you must remove a load-bearing wall, be sure to retain professional advice for providing other support.

The degree of detail presented in this type of drawing is important. The contractor needs to understand what is required, yet does not want to be bothered with details that are common knowledge in the construction industry. In Fig. 14.1, notice the following items. (The note on the drawing is shown first, followed by the contractor's interpretation of what work will be involved.)

1. 4' HIGH CHAIN LINK FENCE & GATE.   The fence will be framed with pipe and anchored to the floor and walls. The gate size and location will be scaled off the drawing.

2. REMOVE THIS WALL.   The wall and doors will be removed, and any holes left in the floor or remaining walls will be patched and finished off.

3. OPENINGS 24" OFF FLOOR 3' × 3'.   Openings will be cut through the walls right next to the outside wall, and all exposed surfaces will be finished off.

4. 3' DOOR.   An opening will be made in the wall at each location shown, and a door and frame will be installed. The door hinges will be located as shown in the drawing. All new doors will have ADA latches and will conform to the minimum opening force.

5. 18" SQ. LUCITE PASS THRU.   The pass-thru will be supplied by the Flexible Division. When it is delivered, a hole will be cut in the new wall, and the pass-thru will be installed and secured with the door on the south side. The pass-thru will then be trimmed with molding on both sides of the wall.

6. 33" WIDE COUNTER.   An opening will be cut in the wall and finished off. A hard-top counter, 33 in wide and 9 in deep, will be mounted on the bottom of the opening. A locking door will be mounted on the inside of the room to cover the entire opening.

The contractor knows that all unshaded walls and all doors will be installed as shown and according to current building code construction specifications. All wall terminations will be finished off to match existing walls.

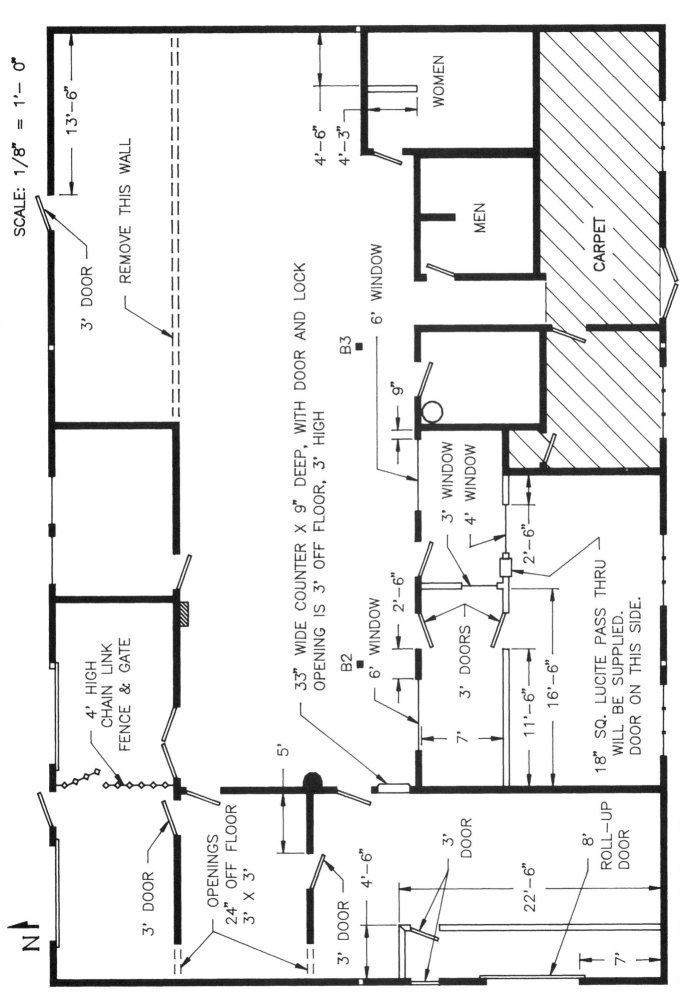

Figure 14.1 Wall modifications.

Notice the column numbers, not really needed on this drawing since there are only two, but they may be an important location aid on your drawing. These column numbers were originally assigned to the new plant in Fig. 8.3.

The notes that accompany Fig. 14.1 are shown in Fig. 14.2. These notes can be shown right on the drawing if they are not too extensive. The notes are an important aid to the contractor because they describe some items not shown on the drawing and add more details to others.

Electrical Services

The next step the IE took to alter the new Flexible Division plant was to create the drawing of electrical services shown in Fig. 14.3. This drawing indicates what power services will be required in the new plant. It is also an example of what you might design as a guide to the general contractor for your new plant modification. The IE examined the existing electrical utilities and then decided what additions would be needed. Only these new utilities are shown on the drawing.

The degree of detail presented in this type of drawing is also important. The electrician needs to understand what service and type of receptacle is required and the location of the receptacle or the power drop. In this case, the receptacle locations are not critical and can be measured right off the drawing.

1. All wall modifications and other work shown on the wall modification drawing and mentioned in these notes will be completed three weeks before moving day, to allow for the utility installation, clean-room HVAC system installation, and drop-ceiling installation.

2. New walls that are to be constructed are shown unshaded.

3. New windows are all 3 ft high and 4 ft from the floor unless otherwise noted.

4. Install conductive floor tiles in all areas unless otherwise noted before any equipment or utility struts are installed. The tiles will be supplied by the company.

5. Paint all walls a light color and all doors and trim an accent color. The colors will be chosen by the company, and the contractor will supply the paint.

6. The roll-up door shown on the west wall is steel, 8 ft wide × 8 feet high, self-powered, with a lock. The complete door will be supplied by the contractor.

7. ESD carpet will be laid on the floor of the front office, which is hatched on the drawing. Do not install tile in this area. The carpet will be supplied and installed by the company.

Figure 14.2 Wall modification notes.

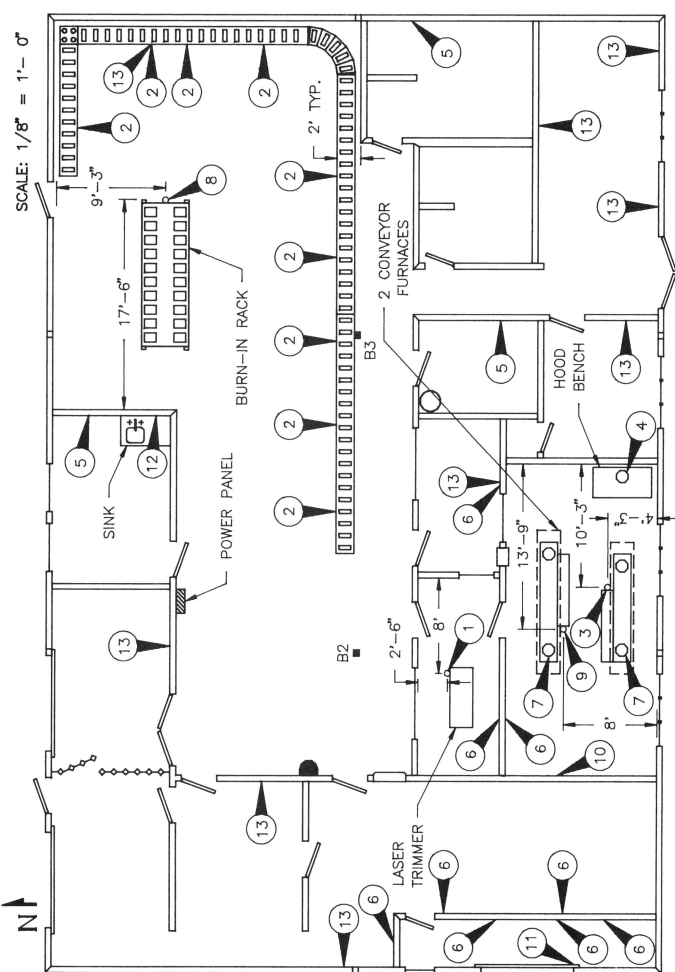

SCALE: 1/8" = 1'- 0"

Figure 14.3 Electrical services.

In Fig. 14.3, observe how the different types of service are indicated. Then notice how each type is described in the electrical service legend in Fig. 14.4. The circle-arrow symbol is designed to identify and reference a detailed description of the type of service needed and to show exactly where it is located (the point of the arrow). This symbol is included in the library of blocks and is named MI-UTIL.DWG.

Electricians normally use standard electrical symbols for this type of drawing, but most people are unfamiliar with them. This circle-arrow symbol and legend combination, however, is easily understood and very usable by everyone. Here is a more detailed description of some of the items shown in Fig. 14.3, for electrical services:

1. *Item 1.* Where a power drop is required, such as for the laser trimmer, its exact distance from the wall is shown on the drawing. Before moving day, the electrician will run the conduit and wire right down to that location, and the wires will extend beyond the end of the conduit.

Before moving day, the laser trimmer will be moved into place. The electrician will trim the conduit to the exact length and will connect the power. When completed, the conduit will be perfectly vertical, and the laser trimmer will be properly located.

To implement this plan, the IE located the exact point of connection of the service drop on the block for the laser trimmer and indicated that location exactly on the electrical service drawing as a guide to the electrician. Brief instructions for the electrician are shown in Fig. 14.4.

2. *Item 2.* The outlets will be placed on struts about 6 in off the floor and will be connected by conduit. On moving day, a bench will be placed next to the conveyor at each outlet.

3. *Item 4.* The power will be connected to the exhaust blower on the roof, but a key-lock switch for the blower will be installed on a wall near the hood bench. If you have power supplied to remote equipment, be sure to specify where the controls should be installed. If the location is not specified, the electrician may mount them near the equipment.

4. *Item 10.* A NEMA L6-20 receptacle will be mounted on the existing wall for the oven. If you have a similar 220-V power outlet on your drawing, be sure to specify the exact type of receptacle you want the electrician to install. Some rare 220-V outlets have a lead time of many weeks and could delay your start-up time if they are not ordered early.

The connector type number will be stamped right on the existing plug and outlet. In an emergency, you can use the outlet from the old plant wall or rewire the plug on the equipment for a more common outlet.

5. *Item 13.* All the Flexible Division computer terminals will be connected to the local area network at corporate headquarters next door. All the system cables will be hung in the overhead above the ceiling tiles, the wires dropped down the wall, and a connector mounted at each location.

If the location dimensions of any utility are not given on the layout, measure the location from the drawing. All new 110-V outlets are specified as ground-fault circuit interrupt with a reset button. After installation, each outlet is to be tested for proper connection.

① *110 V.* Install the power drop for the laser trimmer before it arrives. Connect the power when it is in place. Mount a 110-V quad outlet on the laser trimmer near the power connection.

② *110 V.* Connect power to nine quad outlets facing into the room and mounted about 6 in off the floor on struts at the positions shown, but 18 in from the wall or its extension. Before moving day the conveyor will be installed over these outlets. On moving day, benches will be placed in line with each outlet.

③ *208 V, three-phase, 100 kW.* Install the power drop to the exact location on the new conveyor furnace after it arrives but before moving day. Mount a 110-V quad outlet on the furnace near the power connection.

④ *110 V.* Connect power to the exhaust blower on the roof for the hood bench. Mount a key-lock wall switch for the blower near the hood bench. Complete all work before moving day.

⑤ *110 V.* Connect power to three quad outlets mounted on the existing walls at the locations shown, before moving day.

⑥ *110 V.* Connect power to nine quad outlets mounted in the new walls at the locations shown, before moving day.

⑦ *110 V.* Connect power to two exhaust blowers on the roof for the conveyor furnaces. Mount a key-lock wall switch for each blower near the furnaces. Complete all work before moving day.

⑧ *110 V, 300 A.* Install the power drop to the exact location before moving day. Connect the power to the burn-in rack on moving day after the equipment is in place.

⑨ *208 V, three-phase, 100 kW.* Install the power drop to the exact location before moving day. Connect the power to the old conveyor furnace on moving day after the equipment is in place. Mount a 110-V quad outlet on the furnace near the power connection.

⑩ *220 V.* Connect power for the oven to a 220-V receptacle (NEMA L6-20) on the existing wall before moving day.

⑪ *110 V.* Connect power and wall switch to the roll-up door before moving day.

⑫ *110 V.* Connect power to a duplex outlet about 12 in above the sink counter top before moving day.

Figure 14.4 Electrical service legend.

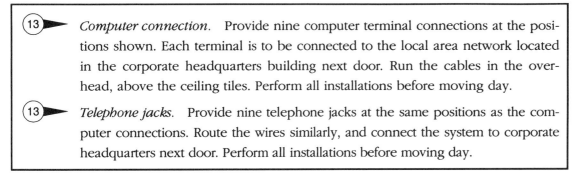

13 ▶ *Computer connection.* Provide nine computer terminal connections at the positions shown. Each terminal is to be connected to the local area network located in the corporate headquarters building next door. Run the cables in the overhead, above the ceiling tiles. Perform all installations before moving day.

13 ▶ *Telephone jacks.* Provide nine telephone jacks at the same positions as the computer connections. Route the wires similarly, and connect the system to corporate headquarters next door. Perform all installations before moving day.

Figure 14.4   Electrical service legend (*Continued*).

Telephone wires and jacks will be installed similarly at each computer connection location. The telephone system will be connected to the corporate system next door.

Notice in the notes at the top of Fig. 14.4 that ground-fault circuit-interrupt 110-V receptacles are specified. This is added insurance that the operators, who are ESD-grounded, will not receive a shock, or worse, from any faulty hand tools. Also, each receptacle is to be tested for proper wiring after the installation is complete. This test, along with a polarity verification, will expose any ungrounded outlets.

The utility location dimensions shown in Fig. 14.3 and other drawings in this chapter were generated by AutoCAD using the DIM command. You can dimension your drawing just as easily, after you place the point of the arrow exactly where you want the utility to be, using a 3-in SNAP distance. Turn SNAP off to place the utility number in the center of the circle.

The IE positioned the exhaust vent and power drop for all the equipment on each block to the nearest 3 in. Then the IE placed the block on the layout to the nearest 3 in, with reference to the columns and walls. Finally, the IE instructed AutoCAD, using the DIM command, to show the distance between the utility and the walls. The distance that AutoCAD specified on the drawing is accurate enough, within 3 in, to be used by the contractor to locate that utility before the equipment is moved into place.

The final hookup is then a simple process that can be done on moving day, when the equipment is moved into position and placed so that the utility drop or ductwork is perfectly vertical. This is the most expedient and visually satisfying way to get the equipment up and running as quickly as possible.

Of course, if you are moving very heavy equipment that must be placed permanently and is not easily adjusted after that, then the utilities should be dropped after placement. However, the utilities can still be installed in the overhead as close as possible to their use and then dropped on moving day. This procedure will require more time than the previous method but will ensure perfectly vertical drops to the machine.

Mechanical Installations

The next step the IE took was to create the drawing shown in Fig. 14.5, showing mechanical installations. This drawing indicates what mechanical installations will be required for the new layout. It is also an example of what you will need to guide the general contractor for your new plant modification.

Greater detail is required in this drawing because this work is less commonplace. It will be performed by plumbers, sheet-metal workers, and mechanical workers who need to understand exactly what is required and where it is located.

In Fig. 14.5, notice how the different types of installations are indicated. Each type is described in more detail in Fig. 14.6. The IE assigned a number series to the mechanical installation items different from the electrical services number series, to avoid confusion when discussing all items with the contractor. Here is a more detailed description of some of the items shown in Fig. 14.5.

1. *Item 20.* The sink and cabinet will be supplied and installed by the contractor. The IE positioned this sink, and also the hood bench, near the outside walls so a minimum amount of concrete would have to be removed for the drain connections.

2. *Items 21 and 41.* The deionized water will be supplied using the reverse-osmosis system from the old plant. The piping will be installed before moving day, the supply system will be moved by the contractor just before moving day, and the hood bench will be connected on moving day. The system will be disinfected and checked out after moving day by the Flexible Division engineers.

3. *Item 29.* The fire extinguishers can be installed on moving day and are mounted with the top 4 ft off the floor so they are accessible to employees who are disabled.

4. *Item 33.* The old conveyor furnace has two exhaust ducts. The one with the dimensions shown is perfectly vertical and exhausts through the building roof. The other joins it with an elbow and tee above the ceiling tiles. Both exhaust ducts will have to be insulated near the bottom to prevent burning the operators. This will be a time-consuming operation for the contractor on moving day and one that will have to be carefully scheduled and cleaned up. The work cannot be done after moving day because the contamination would degrade the air quality and interfere with production.

Notice that on all legends, each installation has a definite time set for completion. The entire project has been carefully scheduled. Some installations will be prepared early and then followed with a brief connection that will be performed on moving day.

New Plant Transformation

There are two alterations to the Flexible Division's new plant that have not been mentioned. One is the HVAC system for the

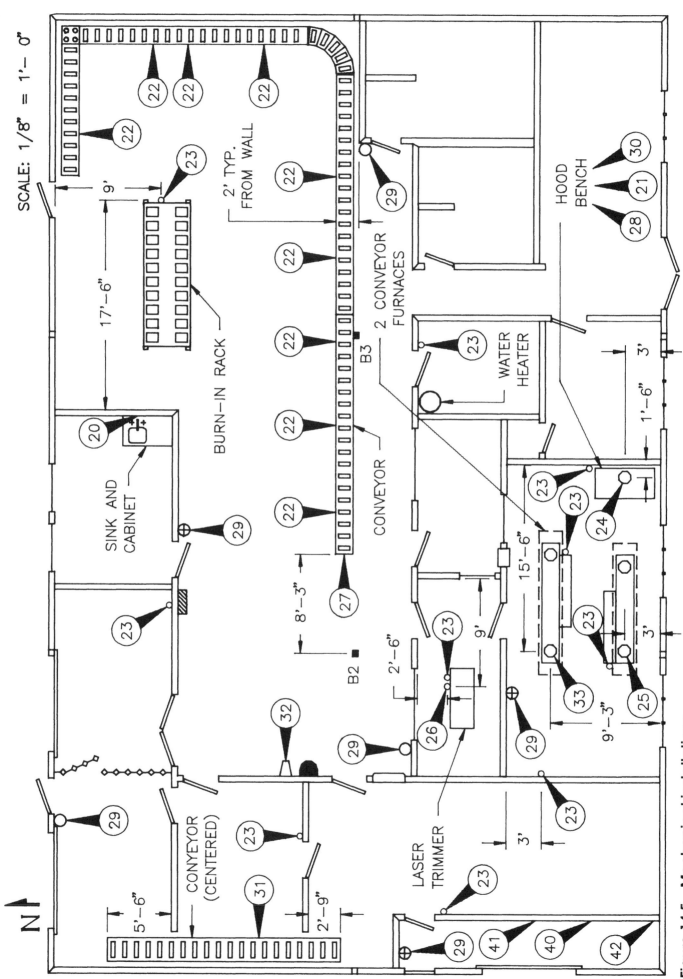

Figure 14.5 Mechanical installations.

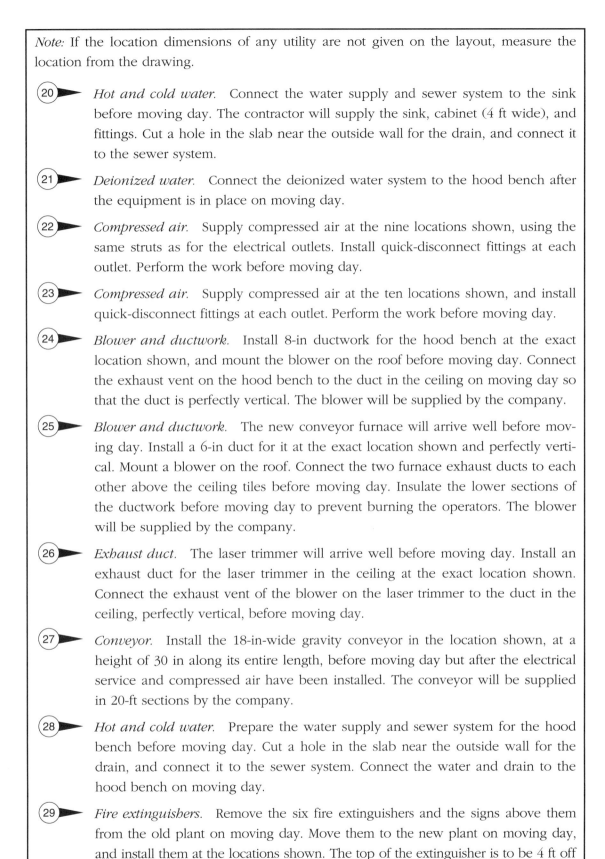

Note: If the location dimensions of any utility are not given on the layout, measure the location from the drawing.

20 ▶ *Hot and cold water.* Connect the water supply and sewer system to the sink before moving day. The contractor will supply the sink, cabinet (4 ft wide), and fittings. Cut a hole in the slab near the outside wall for the drain, and connect it to the sewer system.

21 ▶ *Deionized water.* Connect the deionized water system to the hood bench after the equipment is in place on moving day.

22 ▶ *Compressed air.* Supply compressed air at the nine locations shown, using the same struts as for the electrical outlets. Install quick-disconnect fittings at each outlet. Perform the work before moving day.

23 ▶ *Compressed air.* Supply compressed air at the ten locations shown, and install quick-disconnect fittings at each outlet. Perform the work before moving day.

24 ▶ *Blower and ductwork.* Install 8-in ductwork for the hood bench at the exact location shown, and mount the blower on the roof before moving day. Connect the exhaust vent on the hood bench to the duct in the ceiling on moving day so that the duct is perfectly vertical. The blower will be supplied by the company.

25 ▶ *Blower and ductwork.* The new conveyor furnace will arrive well before moving day. Install a 6-in duct for it at the exact location shown and perfectly vertical. Mount a blower on the roof. Connect the two furnace exhaust ducts to each other above the ceiling tiles before moving day. Insulate the lower sections of the ductwork before moving day to prevent burning the operators. The blower will be supplied by the company.

26 ▶ *Exhaust duct.* The laser trimmer will arrive well before moving day. Install an exhaust duct for the laser trimmer in the ceiling at the exact location shown. Connect the exhaust vent of the blower on the laser trimmer to the duct in the ceiling, perfectly vertical, before moving day.

27 ▶ *Conveyor.* Install the 18-in-wide gravity conveyor in the location shown, at a height of 30 in along its entire length, before moving day but after the electrical service and compressed air have been installed. The conveyor will be supplied in 20-ft sections by the company.

28 ▶ *Hot and cold water.* Prepare the water supply and sewer system for the hood bench before moving day. Cut a hole in the slab near the outside wall for the drain, and connect it to the sewer system. Connect the water and drain to the hood bench on moving day.

29 ▶ *Fire extinguishers.* Remove the six fire extinguishers and the signs above them from the old plant on moving day. Move them to the new plant on moving day, and install them at the locations shown. The top of the extinguisher is to be 4 ft off the floor. An open-circle symbol means water, and a cross means carbon dioxide.

Figure 14.6   Mechanical installation legend.

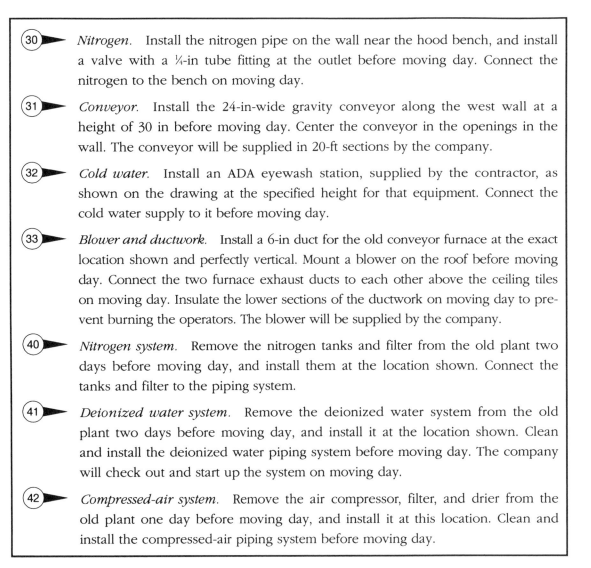

30 ► *Nitrogen.* Install the nitrogen pipe on the wall near the hood bench, and install a valve with a ¼-in tube fitting at the outlet before moving day. Connect the nitrogen to the bench on moving day.

31 ► *Conveyor.* Install the 24-in-wide gravity conveyor along the west wall at a height of 30 in before moving day. Center the conveyor in the openings in the wall. The conveyor will be supplied in 20-ft sections by the company.

32 ► *Cold water.* Install an ADA eyewash station, supplied by the contractor, as shown on the drawing at the specified height for that equipment. Connect the cold water supply to it before moving day.

33 ► *Blower and ductwork.* Install a 6-in duct for the old conveyor furnace at the exact location shown and perfectly vertical. Mount a blower on the roof before moving day. Connect the two furnace exhaust ducts to each other above the ceiling tiles on moving day. Insulate the lower sections of the ductwork on moving day to prevent burning the operators. The blower will be supplied by the company.

40 ► *Nitrogen system.* Remove the nitrogen tanks and filter from the old plant two days before moving day, and install them at the location shown. Connect the tanks and filter to the piping system.

41 ► *Deionized water system.* Remove the deionized water system from the old plant two days before moving day, and install it at the location shown. Clean and install the deionized water piping system before moving day. The company will check out and start up the system on moving day.

42 ► *Compressed-air system.* Remove the air compressor, filter, and drier from the old plant one day before moving day, and install it at this location. Clean and install the compressed-air piping system before moving day.

Figure 14.6 Mechanical installation legend (*Continued*).

thick-film clean room, and the other is the tile ceiling installation in the rest of the new plant.

The thick-film production area is a class 100 clean room with an HVAC system separate from the one in the rest of the plant. A consulting firm, specializing in clean-room design, specified the equipment and filters and then supervised the installation. This system was installed very early in the project and did not interfere with any work described in this chapter.

Originally, there was no internal ceiling in the new plant, and all ductwork and other piping in the high, overhead area was visible from the floor. To create a more appealing workplace, a drop ceiling with lighting fixtures will be installed after the new ductwork and piping are in place, but before moving day. This installation will be done under a separate contract by a firm that specializes in lighted, paneled ceilings.

When you have completed the utility drawings and legends for your new plant, review them with the area supervisors and maintenance manager. As mentioned before, it is imperative that all items needed be specified somewhere in this set of documents. The production supervisors and maintenance manager are generally very familiar with their equipment and can easily spot any utility omissions.

When a piece of equipment has been dismantled for moving, it is relatively easy to modify or improve it. This operation may require extensive planning on everyone's part, but it can save a sizable amount of downtime that may have to be spent later when it is not so convenient. The supervisor and the maintenance manager are the best sources for these suggestions.

This is also a perfect time to paint the equipment. The production area will be shut down for at least one day before the move, and with tight scheduling, the equipment could be painted during that interval. Also, any equipment maintenance that requires downtime should be performed at this time.

The supervisors will play a key role in the move and should be consulted and informed during each step of the plant alteration. Mistakes are easy to make during this phase of the project, and having knowledgeable people review your work for errors is an immense contribution to its success.

When you are sure that the drawings and legends are complete, bidding for the alteration contract can begin. Your company may have experience with one contractor, and a trusting relationship may already be in place. In that case, you can review the project with the contractor and arrive at an agreeable price.

If the contract is up for bid, then the preparation will require more time in the form of general contractor evaluation and price considerations. In either case, you should perform the following steps with each general contractor under consideration:

1. Tour the old and new plants with the contractor.

2. Supply the contractor with copies of the drawings and legends.

3. Discuss each item, and if the contractor suggests that more detail be added to the documents, be sure to include these details in the final versions.

4. If you feel that the contractor may not understand what is needed, have the contractor explain it back to you.

5. Discuss the timing for each installation. Your goal should be to turn on the production area the first working day after the move. If the contractor is hesitant about maintaining that schedule, consider hiring someone else or subcontracting some of the work yourself.

6. While the contractor is preparing the bid, check the contractor's references.

7. The old adage is also true in the construction business: You will probably get what you pay for. The lowest bid may require constant supervising and expediting on your part. Major considerations for qualifying a contractor are quality of work, time available for your job, and a friendly, cooperative attitude.

8. After you choose the contractor, a more detailed installation schedule should be developed between the two of you.

Some information about general contractors' daily problems was given in Chapter 6. Also recognize that subcontractors will be doing much of the work, and you should meet them and their workers. All of them will want to contact you directly if any problems arise.

After the construction is under way, make yourself available on a daily basis by touring the new building and discussing the work with everyone who asks. When several contractors work on a building at the same time, conflicts are a certainty. Many times, you can uncover misunderstandings simply by asking, "How's it going?"

If a major problem should arise, seek the advice of the general contractor or supervisor before directing a worker to change anything. The change could affect the contractor's overall expense and require a revision to the quote for the job.

Your main concern now is to ensure that all the work needed before moving day is completed on time. In Chapter 15, it is assumed that all new plant alterations for the Flexible Division are complete:

1. The conductive tile and ESD carpet have been laid on the floor.

2. The conveyor, power, compressed air, telephone jacks, and computer connections are installed.

3. All the ductwork above the ceiling and extending below it is in place.

4. The water system is installed, the sink in the break room is operational, and taps are available for the hood bench.

5. The clean-room HVAC system is in place and tested.

6. The drop ceiling with the new lighting system has been installed.

7. All interior walls have been painted a light color with an accent color on all trim.

15

Moving Day

In this chapter, a schedule for organizing the move will be created, and the moving company will be selected. The employees will be assigned duties for the move preparation and moving day. The plant layout drawing will be enlarged and used to show equipment locations during the move. Concerning Fig. 6.1, the plant layout project schedule, this chapter will complete tasks 1, 2, 3, 12, 13, 14, and 15.

Move Schedule

The first step the IE made in planning for moving day was to create Fig. 15.1, the move schedule. The WordPerfect file for this schedule, named MOVEPLAN.WP, is included on the diskette and can be used as a model for your schedule. You should create a schedule early in the planning phase for your move. The immense number of events that must take place as part of the move and the reliance on others to perform their duties can make anyone feel apprehensive about the success of the project. Nothing will ease your mind more than having a complete schedule of events and duties that has been reviewed by the important participants and approved for execution.

Start developing the move schedule well ahead of time, even while the plant is being altered by the general contractor. The job list that you assign to the contractor as part of the alteration will, of necessity, contain many tasks directly related to the move schedule. Also, there are so many tasks that must be coordinated during the move that you cannot possibly think of all of them in the short period just before moving day.

The IE started to develop the move schedule right after the alteration contract was awarded. During the alteration review, the general contractor asked about the priority sequence of the

All employees will report for work during normal working hours for the plant move unless additional work is specified. The IE will number all equipment with color-coded labels. Employees should label all boxes of parts or accessories with the same number and color as the equipment they are associated with.

Well before Moving Day

1. Storeroom supervisor

    a. Make a plan for changing the part location sequence in the store-room to subassembly order rather the present part number order. Prepare a list, for the new storeroom, of the shelf rack numbers, the shelf spacing in the racks, and what parts will be stored in them.

2. General contractor

    a. Laser trimmer 711. When it arrives, install the laser trimmer in the new plant.
    b. Conveyor furnace 710. When the new furnace arrives, install it in the new plant.
    c. Vending machines 902 and 903. When they arrive, install both machines in the break room.

Wednesday, Two Days before Moving Day

3. Storeroom personnel

    a. Issue enough parts to production for one day's operation.
    b. Start to pack all parts in boxes for their new shelf rack assign-ments, and use a felt-tip pen to label the boxes.

4. Assembly area personnel

    a. Continue to assemble units, but plan on emptying the burn-in rack by moving-day morning. Move all finished units that are not burned in to the conveyor in the new building.

5. Thick-film room personnel

    a. Conveyor furnace 704. Shut down the furnace at the end of the day.
    b. Batch oven 707. Turn off and unplug the oven at the end of the day.

6. All personnel

    a. Packing. The packing boxes will be supplied by the mover and will be stored in the utility room on Wednesday. All employees will pack the belongings for their work stations on Thursday.
    b. The IE and the move coordinators will tag all equipment with self-stick labels that are color-coded for the work center (assem-bly, storeroom etc.). The equipment number will be on the label.

Figure 15.1   Move schedule.

7. Mover

    a. Deliver folded packing boxes to the utility room. Tour the old plant, look for problem areas, and report them to the IE.

8. General contractor

    a. Deionizer. Remove the deionizer from the old plant and install it in the new plant.

    b. Nitrogen tanks. Remove the nitrogen tanks and filters from the old plant and install them in the new plant.

    c. Thick-film room. In the new building, turn on the ventilating system, and clean the walls according to the standard instructions provided.

Thursday, One Day before Moving Day

9. Trash

    a. Deposit trash in the dumpster behind the old plant.

10. Storeroom personnel

    a. Finish packing all parts, and use a felt-tip pen to label the boxes for their new shelf rack assignments. Enlist the help of other employees when they are through packing their belongings.

    b. Disassemble the shelf racks and pack each one in a tied bundle. After each shelf rack is tied up, be sure that the equipment number label is still plainly visible on the bundle.

11. Thick-film room personnel

    a. Batch oven 707. When the oven is cool, clean the inside and outside as well as possible.

    b. Hood bench 701. After the contractor has dismantled the bench, pack all the parts in boxes and label them. Label the exhaust duct also.

    c. Screen printer 706. Remove all loose parts, tie down the moving parts, and prepare the printer for moving.

    d. Deionizer. In the new building, engineers from corporate headquarters will decontaminate the deionized water system and check it out completely.

12. All personnel

    a. Do not remove any equipment labels. When an area is fully packed and ready to be moved, the IE and the move coordinators will verify all labels.

    b. Pack the belongings for your work station in boxes, and use a felt-tip pen to label each one with the number of the equipment it goes with. Label your wastebasket with the same number.

    c. The desks and workbenches will be moved in the upright position, so items may be left in the drawers, but lock or tie them shut. Do not leave any loose items on top of the workbenches.

Figure 15.1   Move schedule (*Continued*).

> d. Help pack parts in the storeroom, and help prepare equipment in the thick-film room when you are through packing your own belongings.
>
> e. Copies of the new plant layout will be posted in the new and old plants.

13. Mover

> a. *Old and new plants.* Cover the carpeted, high-traffic areas with hardboard. Install door jamb protectors on all high-traffic doors. Remove the doors and protect the door jambs leading to the thick-film room.

14. General contractor

> a. *Conveyor furnace 704.* When the furnace is cool, disconnect the power, remove the covers, clean and vacuum the internal parts as well as possible, and clean the outside covers. Replace the covers.
>
> b. *Air compressor.* Remove the air compressor and filters from the old plant and install them in the new plant. Connect them to the compressed-air piping system.
>
> c. *Hood bench 701.* Disconnect all services to the bench. Disconnect the exhaust duct and leave it near the bench.

Friday, Moving Day

15. Trash and empty boxes

> a. Deposit trash in the two dumpsters behind the new plant. Break down all empty cartons that belong to the mover, and stack them in the receiving area.

16. No-tag zone

> a. In the new plant, place all items that do not have labels in the staging area near column B2. All employees should make every effort to identify no-tag items and move them to the proper location.

17. Coordinators

> a. The moving coordinator for the old plant will be the Quality Assurance Supervisor, and the coordinator for the new plant will be the Production Supervisor. They will be in charge of duty assignments for all employees on moving day.

18. Assembly area personnel

> a. The technician should load units into the burn-in rack after it is connected and powered up. Burn-in cycle times for the units may be continued from where they left off in the old plant.

19. Thick-film room personnel

> a. In the new building, unpack the parts for the hood bench, and assist the contractor in reassembling it. Connect the washer to the deionized water supply.

Figure 15.1   Move schedule (*Continued*).

20. All personnel

    a. Report to the new plant unless the coordinators assign you to other duties.

    b. Familiarize yourself with the new plant layout drawing that is posted in many locations in the new plant. The faded lines on the layout are 1 ft apart, and all equipment is located to the nearest 3 in.

    c. When the equipment arrives in your area, show the movers exactly where it goes. The coordinator will supply each area with a tape measure.

    d. Unpack all boxes, and assemble each work station as shown in the new layout. Remove all labels after the equipment is in place.

    e. If extra time is available, assist the storekeeper in unpacking and loading the parts into the racks in the new subassembly order. Assist the thick-film room personnel in starting up their equipment.

21. Mover (generally, move items in the order listed)

    a. Burn-in rack 123. The rack should be empty. The electrician will disconnect the power to the rack. Move the rack as one unit. Place it in location in the new plant so the power drop is perfectly vertical when it is connected.

    b. Hood bench 701. Move the bench and all fittings, including the exhaust duct, to the new plant.

    c. Thick-film furnace 704. Move the furnace to the new plant, and place it in location so the power drop is perfectly vertical when it is connected.

    d. Storeroom (300 series)

    e. Assembly area (100 series)

    f. Thick-film room (700 series)

    g. Office area (500 series)

    h. Office area (200 series)

    i. QA area (400 series)

    j. Shipping and receiving (600 series)

    k. Other areas (800 and 900 series)

22. General contractor, old building

    a. Burn-in rack 123. Disconnect the power to the burn-in rack.

    b. Fire extinguishers 1-6. Remove the six fire extinguishers and signs from the old building, and move them to the new building.

23. General contractor, new building

    a. Hood bench 701. Connect the exhaust duct system to the hood, perfectly vertical. Connect the hot and cold water, deionized water, and nitrogen to the bench as they were connected in the old plant.

    b. Burn-in rack 123. Connect the power drop to the burn-in rack, perfectly vertical.

Figure 15.1 Move schedule (*Continued*).

> c. *Conveyor furnace 704.* Connect the power drop to the furnace. Install insulation around the exhaust ductwork.
>
> d. *Fire extinguishers 1-6.* Mount the six fire extinguishers and signs as shown in the layout.
>
> **Saturday and Sunday, after Moving Day**
>
> 24. Storeroom personnel
>
>> a. If necessary, continue to organize the storeroom until the new parts are all in the new location sequence. The storeroom should be ready to issue kits on Monday morning.
>
> 25. Thick-film room personnel
>
>> a. *Conveyor furnaces 704 and 710.* After the furnace insulation is installed and the room is cleaned, wipe down all equipment with a cleaning solution and lint-free cloths. Fire up both conveyor furnaces, and profile them after they have stabilized.
>>
>> b. *Batch oven 707.* Turn on the oven, and adjust it to the proper temperature.
>>
>> c. *Hood bench 701.* Turn on the exhaust blower, and set the damper for the proper airflow in the hood opening. Check all services for proper operation, and turn off the blower.
>>
>> d. *Ventilation system.* Turn on the ventilation blower, and check the room temperature.
>
> 26. General contractor
>
>> a. *Thick-film room.* Clean the room walls with cleaning solution and lint-free cloths. Vacuum and mop the floor of the room.

Figure 15.1   Move schedule (*Continued*).

various jobs. Most job priorities were covered in the utility drawing notes shown in Chapter 14. However, many more jobs would have to be completed in conjunction with moving day. The contractor needed a detailed schedule of these jobs so they would dovetail with the alteration work that was completed earlier.

The move schedule formed the basis for integrating that alteration work with the other work associated with moving day. Another important function of the move schedule is to assign duties to every employee in the plant. Without it, no one would know what to do, and chaos would be inevitable.

As you can see by the schedule for the Flexible Division in Fig. 15.1, the contractor, the mover, and the employees must be closely coordinated before large pieces of equipment can be relocated in the new plant. The burn-in rack is a good example: First it is emptied of burn-in units by the test personnel, then disconnected by the electrician, next moved by the mover, then connected by the electrician, and finally loaded with units to be burned in by the test personnel.

This was all accomplished the first thing on moving day, to minimize the downtime of this critical equipment. Otherwise, it would have dragged out over the weekend. Even so, the weekend was still needed for other work that could not be done on moving day.

Figure 15.1 is more than a schedule for the move and a list of duties to be performed by all persons involved. It identifies by number the important equipment that needs special handling. Examine this schedule closely. The IE developed it over many weeks to ensure that every category was included:

1. *Information.* Everyone will have a copy of the move schedule and will therefore see the whole picture of what has to be done and who will do it. If someone has to assume some other duty on moving day, he or she will immediately know how that duty fits into the overall strategy.

2. *Employees.* Duties are given for the employees of every work center. If they complete their assignments, there are directions for helping others who are shorthanded.

3. *Mover.* The mover has definite work assignments that start two days before moving day.

4. *General contractor.* The contractor has definite work assignments every day of the move. They are mainly extensions of the plant alteration work that the contractor has accomplished just before the move.

5. *New equipment.* The thick-film operators and Chris, the engineer, have eagerly awaited the laser trimmer and conveyor furnace for months, and now they know when and how these will be installed.

6. *Storeroom.* The storeroom supervisor, Pat, has been organizing the new-parts storage sequence for some time and now must formalize that plan. Other employees will help pack and unpack the parts for the storeroom. In fact, this is the major portion of the labor involved in the move.

7. *Burn-in rack.* This equipment will be attended to in some way on every day of the move. Gerry, the production supervisor, will coordinate the duties of each group to ensure an expeditious move to the new plant.

8. *Thick-film room.* The room and equipment will be cleaned before and after the move. All equipment and utilities will be turned on and adjusted so the facility will be ready for operation on Monday morning.

9. *Equipment labels.* Instructions for labeling the equipment are provided. The IE and the coordinators are directed to verify the labels before moving day. On moving day, the labels will guide the movers to the right location for every box and piece of equipment. The employees will then show the movers exactly where to place the equipment.

10. *Move priority.* The mover is directed to move certain equipment first and then to move the work center areas in order of importance. This ensures that the equipment most important for continuing production (the burn-in rack) is moved early and is immediately loaded with units for the burn-in cycle. The hood bench and conveyor furnace have priority because extensive hookup time is required.

The move schedule is a crucial requirement toward meeting one of the most important goals of the move: starting production on the next working day. While you are creating the schedule, imagine what will be required the very next morning to start production. Then plan everything to achieve that condition. Review the schedule with the move coordinators, general contractor, mover, and anyone else who is interested. Make sure that everyone realizes what the goals are, use their suggestions, and the move will be a surprising success.

The Flexible Division was actually able to start assembly on a modest scale on the afternoon of moving day. Because the utilities were installed at each bench location, the operators unpacked their tools, plugged in their equipment and air hoses, and started assembling units right where they had left off the day before.

Mover Selection

The best source for recommending a mover is a company like yours that has moved recently and was satisfied with the job. If you know of one, call and discuss it with the person in charge. Also ask the people in your plant, especially the maintenance personnel, if they have had experience with local movers. Outside of finding a direct recommendation, the Yellow Pages are the best guide under the category *Movers*. Consider only those companies that use the term *industrial movers* in their advertisements.

Review Fig. 15.2, the mover selection checklist, before you interview any representatives of moving companies. This list is a good start toward qualifying a mover, but it cannot completely describe every situation that may arise during a plant move. Examine the equipment in the old plant yourself, and look for situations that the mover may have problems with.

When you interview movers, the representative will want to tour the old and new plants and study your new plant layout. The degree of organization will have a direct bearing on the price that the mover will quote for the job. The mover knows that it will take longer to place the equipment in the new plant, at a higher price, if there is no exact plan for its location.

To make the plan more clear, the IE made color plots of the old plant with equipment numbers (Fig. 10.4) and the new plant with equipment numbers (Fig. 12.2). Before making the tour, the IE explained the move plan and supplied a copy of the move schedule along with both plant layouts to the representative. This combination of documents made a convincing demonstration of the project's organization. The degree of preparation and the planned participation of the employees ensured that the representative would submit a minimum bid for the job.

While the IE was touring the old plant with the first mover, it became apparent that some of the equipment would have to be

Some of the items listed here are samples of the many subjects that you will discuss with the prospective mover during the interview. Make notes of any agreements made during this discussion and formalize them into an addendum that will be part of the final moving contract.

1. *Mover selection.* Survey other people, in your company or in other companies, who have had experience with local movers. Almost every company has moved at one time or another, and the people in charge of that move will have definite opinions about the mover. Ask the mover for references—preferably companies that have been moved recently.

2. *Mover interview.* Have your version of Fig. 15.1, the move schedule; Fig. 10.4, the old plant with equipment numbers; and Fig. 12.2, the new plant with equipment numbers, available to indicate the size of the job and to determine if the mover is qualified to do it. Use your version of Fig. 15.3, the equipment to be moved, if it is needed.

3. *Numbering system.* Explain how all equipment will be labeled, color-coded, and numbered before moving day and how employees will be stationed as guides at each color-coded area in the new plant.

4. *Moving special equipment.* Inquire about the methods used to prepare certain equipment for the move:
   a. *Desks.* Will the desks be moved in the upright position? If not, can the drawer contents be clamped in some manner and the drawers secured so the desks can be moved with the drawers inside? Will the top center drawers still have to be moved separately?
   b. *File cabinets.* Can the files in the drawers be clamped and the drawers secured in the cabinet so they can be moved as a unit?
   c. *Workbenches.* Will they have to be dismantled in order to pack more efficiently? If so, who will do the work?
   d. *Storage racks.* Some racks snap apart easily while others are assembled with many bolts. How will each type be prepared for moving, and who will do it?
   e. *Pallet racks.* Can the pallets be moved without repacking? Who will dismantle the pallet racks?
   f. *Computers.* Will padded containers be supplied? Who will do the packing and unpacking?
   g. *Bookcases.* Will the books be packed separately, and who will do it? Or can they be strapped into the bookcase and moved as one unit?

5. *Moving heavy equipment.* If some of your equipment is so large that it must be separated before moving, who will perform the work? Sometimes it is better to have sensitive electronic equipment dismantled by the engineers who operate it. Does the mover have special tools for manipulating heavy pieces of machinery? Long pry bars, industrial dollies, A-frame hoists, and heavy fork lifts are often needed. Have the mover describe the exact procedure that will be used for pickup and delivery.

Figure 15.2 Mover selection checklist.

6. *Plant protection features.* The carpets in the old and new buildings must be protected during the move with hardboard runners, supplied by the mover. The mover should also mount door jamb protectors on all main-traffic doors to prevent almost certain damage.

7. *No-tag zone.* Discuss the location of an area in the new plant where unlabeled equipment can be stored until it is identified.

8. *Box area.* Discuss the location of an area in the new plant where the mover's boxes can be placed when they are empty.

9. *Trash area.* Discuss the location of an area and dumpster outside the new plant where trash can be disposed of.

10. *Emergencies.* For some strange reason, truck maintenance seems to be an obscure concept to many short-haul moving companies. Their trucks are used constantly and often break down at a crucial time in the move. Ask the mover how many trucks will be required and what will happen if one of them breaks down. It is generally wise to plan on the move taking half again as long to complete as the mover says it will.

11. *Insurance.* Be sure the mover has liability insurance to protect against serious damage to your company's buildings, equipment and employees. The mover should also be covered by Workers' Compensation to eliminate the possibility of your company's paying for the mover's personnel accidents.

Figure 15.2 Mover selection checklist (*Continued*).

moved by the Flexible Division. The IE was constantly crossing out these items on the layout. To clarify this situation, the IE made a modified version of the new plant layout to include only the equipment that would be moved by the mover. This drawing is shown in Fig. 15.3.

By reviewing this drawing, the mover could plan precisely what equipment and containers would be involved in the move. No safety margin would have to be included to cover omissions because all the equipment was well defined. If you have a similar situation, with many pieces of equipment in the old plant that the movers will not handle, make up your version of Fig. 15.3 as a guide to the mover for figuring the estimated cost.

Moving Day Preparation

After the move schedule has been completed and the mover has been selected, there are a few more items to consider before the moving process begins.

Aisle Tape

One task that should be performed before moving day is usually forgotten until later, when much of its value has been lost. That task is the application of the aisle-marking tape. The aisle-marking tape is a convenient guide to the placement of equipment on

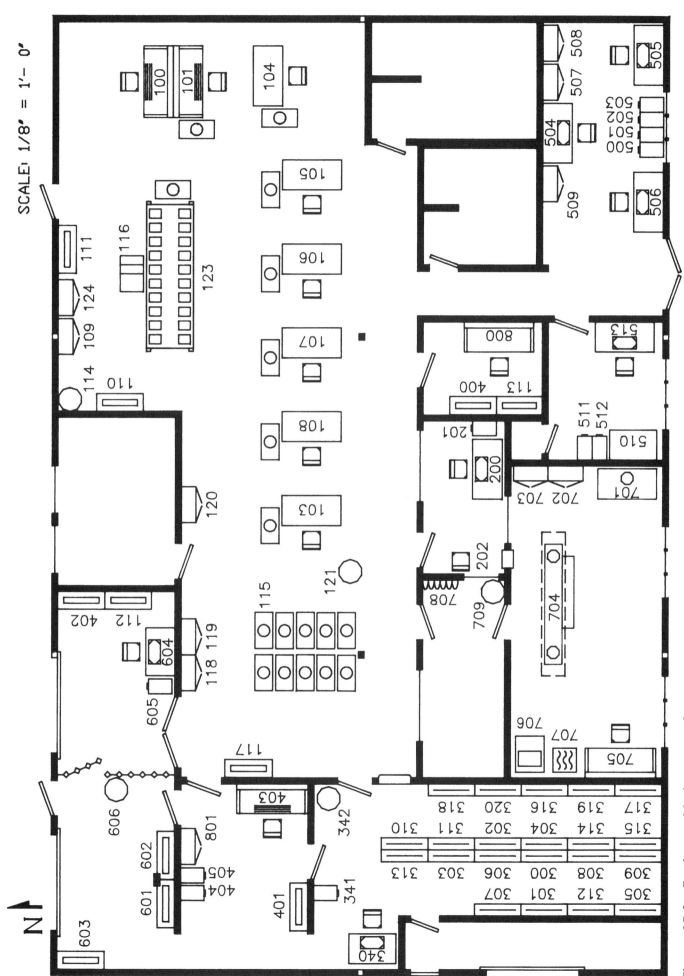

SCALE: 1/8" = 1'- 0'

Figure 15.3 Equipment to be moved.

moving day since many of the items on the layout are located relative to the aisle line.

For instance, when the movers wheel a bench into the assembly area and the bench will be located along an aisle, they can first roll it over to the aisle tape and then measure off to a wall or column for final location. The placement of a row of benches becomes a simple routine, and the aisle tape lines everything up perfectly.

The tape itself is available from your material handling equipment supplier. It comes in a variety of colors and should be chosen to complement the color of the flooring. Do not worry about choosing a color that will stand out; any color tape will be visible once it is on the floor. It is better to choose a color that is a slightly darker accent to the floor color. Shades of brown or gray are popular.

While you are ordering the aisle tape, you can also buy other marking items from the same vendor:

1. *Safety hazard tape (yellow/black).* This is used to denote access areas for power panels, fire extinguishers, water valves, wall ladders, indoor parking areas, or other areas that must be kept clear at all times.

2. *Aisle-marking dots.* Dots can be used instead of tape to delineate areas that have a distinct classification such as a high-security area.

3. *Traffic arrows.* They can be used to direct traffic down one-way aisles or to guide plant tours through the correct route.

4. *Stencil kits.* These are composed of letters and arrows and can be used to direct traffic in large floor areas or parking lots.

5. *Signs.* These can be used to mark many conditions: FIRE EXTINGUISHER (required), EXIT (required), AUTHORIZED PERSONNEL ONLY, DANGER KEEP OUT, CAUTION: SAFETY GLASSES MUST BE WORN AT ALL TIMES and many more.

When you buy tape, choose a supplier who will also lend you an aisle tape applicator or rent it to you at a small cost. Place the tape just outside the aisle line shown on the layout. In other words, the total aisle width should be free of tape. When you apply the tape with an applicator, move very slowly and guide it in as straight a line as possible. Even with the utmost care, it is difficult to achieve a perfectly straight line. When you sight down the line from one end, it may look wavy, but rest assured, this is the last time anyone will examine it that closely.

The Flexible Division required only a small amount of tape to mark the aisles and did not need an applicator. Also, the supplier required a large amount of tape for a minimum order, so the IE bought a roll of light brown vinyl plastic tape at the hardware store and applied it by hand. To complete the task, the IE bought some signs from the supplier to mark the exits and safety-glass area and had the contractor mount them on the wall.

Plant Layout Drawings

Moving day will be the unveiling of your masterpiece, the new plant layout. It is not the exact layout that you have used so far, however, but a larger-sized drawing that can be read on the fly as movers and other personnel pass by with a box of books in their hands. This layout is drawn at a scale of $\frac{1}{4}$ in = 1 ft and may have to be divided into work centers and plotted on several sheets of E-size paper if your plant is huge.

The IE was able to plot the entire layout on one sheet of C-size paper, which is too large to include in this book, but a small section is shown in Fig. 15.4. The dots are spaced 1 ft apart, and nothing has been sized down to suit the larger scale. That is, the text is double-height. If you have a legend on your drawing, do not change the height to suit the new scale. This drawing is made large on purpose, so it can be laid on the floor and read by people standing over it.

The dot pattern is formed by an array of points that coincide with the grid dots on the computer screen when the grid spacing is set at 12 in. Notice that one dot coincides with the center of column B3. In that way, the equipment location can be easily determined by counting the dots (feet) from the center of the columns. For instance, the front of bench 106 should be 5 ft 3 in from the column center. That is one reason why the columns were placed at even-foot locations on the blank layout drawing before anything else was drawn.

An alternative to grid dots is the grid lines shown in Fig. 15.5. Since many copies of the layout will be needed on moving day, it is usual to plot the master on vellum and make copies on a blue-line reproduction machine. The grid lines can be reproduced in fade-out blue if they are plotted on vellum with a blue pen. Almost any shade of blue will work, but all other colors reproduce as though they were black.

Try some pens of various widths along with a blue pen on a scrap of paper to see what combination is best for your large-scale drawing. If you make the grid lines with a narrow blue pen and all other lines with a wide black pen, the features of the layout will be easy to see and the grid lines will be barely noticeable.

With some testing, you will find that either grid dots or grid lines will be better for your application. One factor may influence your choice—when you plot the grid dots, it will sound like a machine gun is leveled at your plotter.

This large-scale drawing is used to locate all equipment on moving day. As the equipment is moved into place, the distances to aisle markers, columns, and walls can easily be determined by counting the 1-ft squares on the layout and then adding on the 3-in increments. Everyone learns how to measure by this method and, by using a tape measure, can place the equipment easily.

There is one refinement to this drawing that is rarely used but is the height of convenience if it can be incorporated into the

NOTES: 1. GRID DOTS ARE 1 FOOT APART.
2. ALL EQUIPMENT IS LOCATED TO THE NEAREST 3 INCHES.

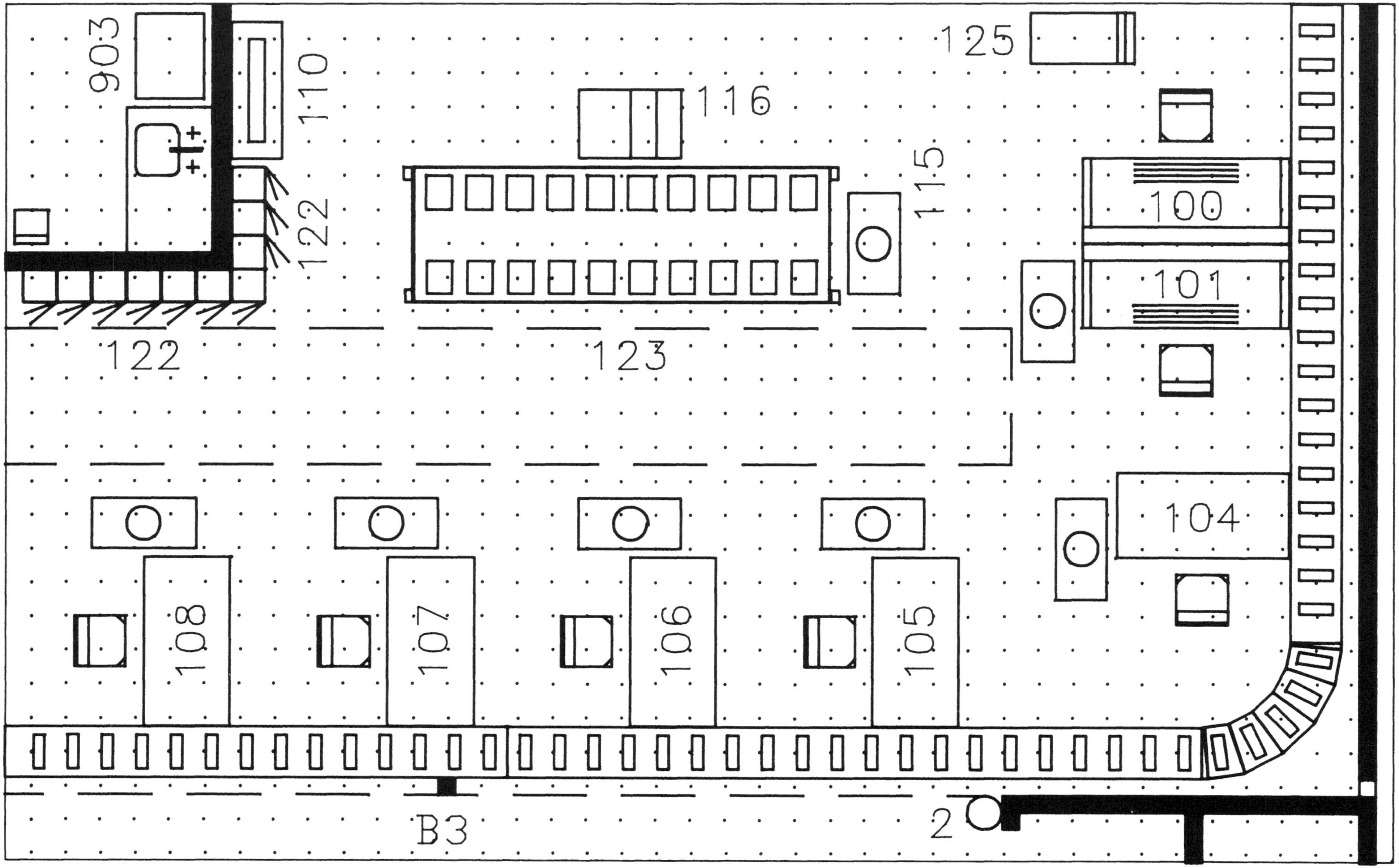

SCALE: 1/4" = 1'— 0"

Figure 15.4 Equipment location grid dots.

NOTES: 1. GRID LINES ARE 1 FOOT APART.
2. ALL EQUIPMENT IS LOCATED TO THE NEAREST 3 INCHES.

SCALE: 1/4" = 1'- 0"

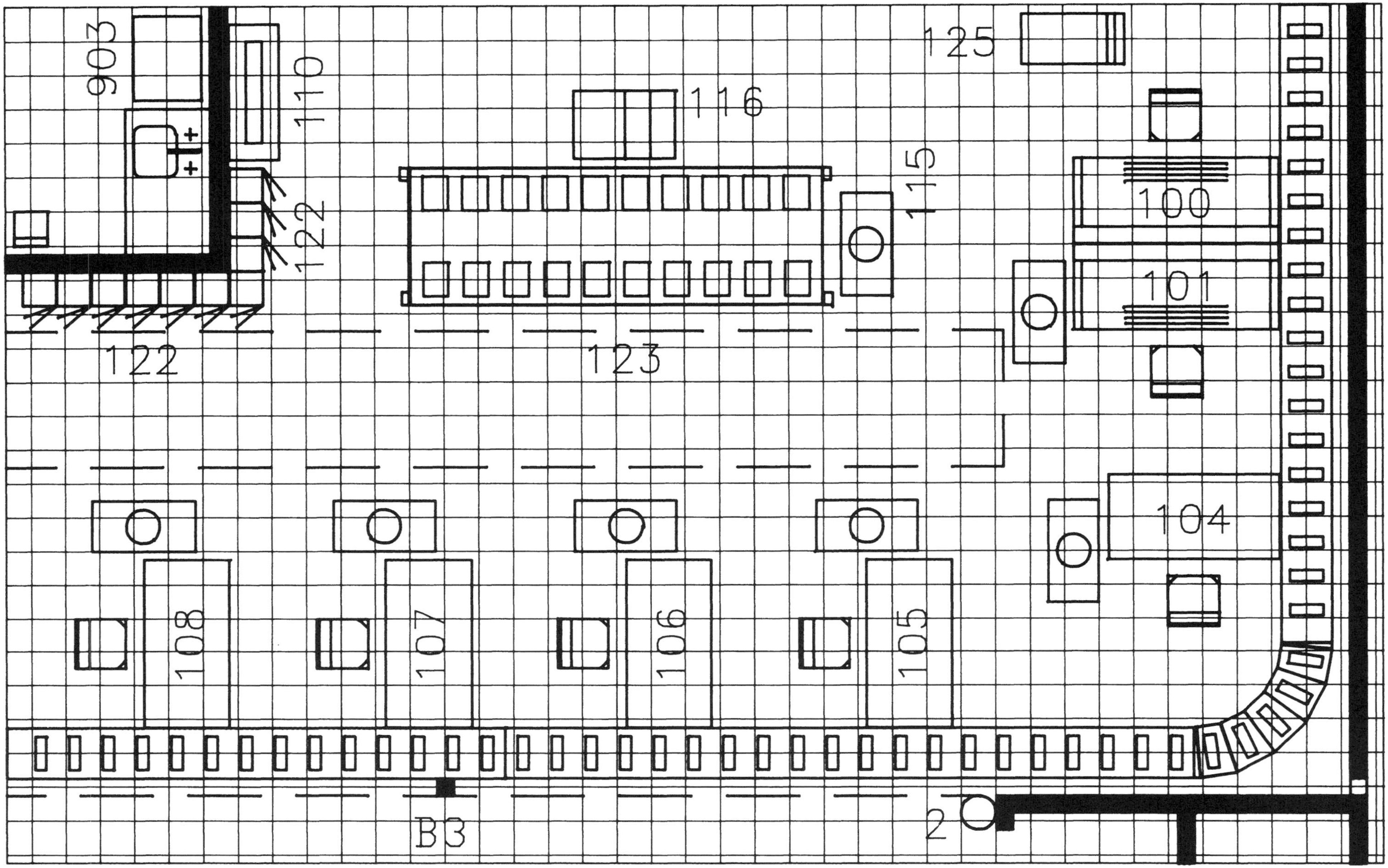

Figure 15.5 Equipment location grid lines.

large-scale layout drawing: If the floor tiles are 1 foot square and are laid down tight together, the grid lines on the layout can be positioned to coincide with the floor tiles, to the nearest 3 in. Before you perform this feat, measure the overall dimensions of the room and count the tiles to be certain that they agree. Then, on moving day, since the grid lines on the layout represent the tiles on the plant floor, equipment location is amazingly simple. Tape measurements are not needed at all.

Coordinators

You should review the move schedule, duty assignments, and any questions with the move coordinators well before moving day. By moving day, they should be as familiar with the moving company personnel and the new plant layout as you are. Everyone involved with the move should know that the equipment labels will be color-coded for each work center and that the equipment number on the label corresponds to the number on the layout.

Moving Day

This is the day that can be smooth and rewarding if it is well organized or utter chaos if it is not. Your anxiety level will be inversely proportional to the amount of planning that you have done. If time was not available to check out the layout and hasty preparations were made for moving day, your anxiety level will be high, and with good reason. Small mistakes made on the layout can escalate to catastrophes on moving day.

As an example, suppose you planned many rows of pallet racks to start at a wall and end at a 10-ft aisle. You assumed that a 10-ft pallet rack was 10 feet long, and you did not have time to check the actual length, which was 3 in longer. On moving day, the movers place the racks, and the last rack extends 2 feet out into the aisle. Rather than order new 8-ft racks and wait several weeks for delivery, the materials manager decides to rearrange the racks on the spot. The result is a haphazard layout with narrow aisles and very inefficient space utilization. In addition, the whole episode is a source of irritation between you and the manager from that time on.

If you have adequate time for the project and plan everything carefully, the move should be very routine without major problems. Your duties on moving day should consist of overseeing the project and ensuring that the general contractor connects all the equipment on schedule. The Flexible Division IE spent most of the moving day working with the plumbers and electricians, sighting in the vertical pipes and conduits.

There is bound to be some work required after moving day that may or may not have been anticipated. That is why the Flexible Division moved on Friday and reserved the weekend to take care of the final details. The thick-film operators came in on Saturday to clean up the room and turn everything on after the mechanics installed the exhaust duct insulation.

Pat, the storeroom supervisor, spent the whole weekend checking out the arrangement of parts stored on the shelves in subassembly order and making new pull lists. It soon became obvious that the kits could be loaded in less time that it took before the move.

On Monday morning, the plant was in full operation and required only minor adjustments to the new layout. The new assembly procedure had been tried out on a prototype basis in the old plant, and it performed even better than expected.

The plant manager was so pleased with the performance of the employees and the success of the move that the manager arranged to have lunch catered for everyone. Even the painters, who were still touching up, were invited. A layer cake was the finale, and the IE was nominated to cut the first piece.

The IE was subsequently moved to corporate headquarters, next door, to head up a team of industrial engineers who would supervise all divisional plant layout and flow improvement projects for the corporation.

The entire plant layout project, including the vendor qualification program, took about a year to complete, but the results were impressive—so impressive that the Flash Memory Bank would dominate the memory field for the foreseeable future. The 4Q vendors were assigned even more products to supply to the Flexible Division, and the qualification process for other vendors became a matter of routine.

An incident occurred about a month after moving day that emphasizes the accommodation of those with disabilities. Terry was alone in the plant during lunch break, as the others had all left on personal business. The thermostat on an assembly soldering iron failed. The iron overheated and started a fire on one of the assembly benches. Terry immediately came out of the break room, lifted the fire extinguisher off the wall, wheeled over to the fire and put it out. When the others returned to work, they were overwhelmed with gratitude. Terry was given the afternoon off for saving everyone's job, and the IE was congratulated for recognizing the need for a lower fire extinguisher installation.

Conclusion

Several concepts that have been emphasized in this book are summarized under these headings:

Layout Benefits

It is a lost opportunity when you make a layout merely to move a production system from one location to another. A new plant layout offers the opportunity to implement a new production system capable of improving the product quality, decreasing the product cost, and reducing the flow time. By all means, expand the scope of the layout project to include some form of product improvement.

Company Support

Any project of this magnitude requires dedication by the entire work force, managers through operators. You have probably read so many similar statements in magazine articles that, by now, it seems more than obvious. Nevertheless, it is the starting point and must be satisfied before you, as the artist, set up your easel.

Problem Solving

Don't treat the symptoms; solve the problem. The latter approach may be more difficult since the symptoms are obvious, while their cause is usually well hidden. Concentrate on the series of events that created the symptoms until you uncover the ultimate cause. The cause of the problem may be more difficult to correct than the symptoms, but when you succeed, you will eliminate the symptoms forever.

Benefits of CAD

If you are not trained in using a CAD program, I hope this book will encourage you to start. A CAD program can provide solutions to a host of your problems in addition to flow charts and plant layouts. The speed and quality of output cannot be matched by any other method.

A CAD program can allow you to quickly modify a file and create a new version of an original layout. Fig. 15.3, showing the equipment to be moved, for example, was created in only a few minutes from the new plant layout drawing file. After the move, the IE used the new plant layout to make an evacuation route drawing simply by turning off the equipment layers, adding the personnel flow arrows, and changing the title.

Simplicity

Simplicity is a virtue in the manufacturing industry. The computer can be the antithesis of simplicity, and the temptation to use it to develop complex solutions is very strong. Regarding your layout, by all means keep it simple with straightforward flow patterns and uncomplicated equipment templates. Any time you see a very elaborate and detailed layout, you should recognize that its creator likes to waste time.

Regarding other system problems, imagine that the computer has not been invented. Solve the problem by using simple techniques, then add the computer for speed and reliability.

Final Remarks

With the help of the Flexible Division IE's example, you can now approach the task of laying out your new plant with confidence. Your new plant layout will be a very real asset to your company, and your expertise in this field will expand greatly. Good luck on your project—I hope you enjoy the challenge, and find that your work has been a rewarding experience.

References

1. Gavriel Salvendy, (ed.), *Handbook of Industrial Engineering,* 2d ed. Copyright © 1992 and published by John Wiley & Sons, Inc., New York All quotations are reprinted by permission of John Wiley & Sons, Inc.

2. Institute of Industrial Engineers, *Industrial Engineering Terminology*, rev. ed., Copyright © 1991 by Industrial Engineering and Management Press, McGraw-Hill, New York, and the Industrial Engineering and Management Press, Norcross, GA.

3. AutoCAD, computer-aided design program, and *The AutoCAD Resource Guide* (packaged with the AutoCAD program). Available from Autodesk, Inc., 2320 Marinship Way, Sausalito, CA 94965. Phone information: 1-800-445-5415.

4. Generic CADD, computer-aided design program. Available from Autodesk Retail Products, 11911 N. Creek Parkway South, Bothell, WA 98011. Phone information: 1-800-228-3601.

5. LOTUS 1-2-3, computer spreadsheet program. Available from Lotus Development Corp., 55 Cambridge Parkway, Cambridge, MA 02142. Phone information: 1-800-343-5414.

6. WordPerfect, computer word-processing program. Available from WordPerfect Corporation, 1555 N. Technology Way, Orem, UT 84057. Phone information: 1-800-451-5151.

7. Eliyahu M. Goldratt and Jeff Cox, *The Goal: A Process of Ongoing Improvement*, 2d ed., paperback edition. Copyright © 1992 by Eli Goldratt, North River Press, Inc., Croton-on-Hudson, NY.

8. Quick Schedule Plus, project scheduling program. Available from Power Up! Software Corp., P.O. Box 7600, San Mateo, CA 94403-7600. Phone information: 1-800-851-2917.

9. CADKEY, computer-aided design program. Available from Cadkey, Inc., 4 Griffith Rd. North, Windsor, CT 06095-1511. Phone information: 1-800-394-2231.

10. Ellen Sugarman, *Warning: The Electricity Around You May Be Hazardous to Your Health*, paperback. Copyright © 1992 by Ellen Sugarman, Simon & Schuster, New York.

11. Tape recommended for hinge on digitizer overlay cover. Manufacturer: 3M Corporation, St. Paul, MN 55144. Stock number: Scotch ® Brand Tape, Cat. 148. Dimensions: 2 in wide × 500 in long. Phone information: 1-800-342-7561.

12. *Nondiscrimination on the Basis of Disability by Public Accommodations and in Commercial Facilities; Final Rule* (ADAAG—ADA Accessibility Guidelines). Federal Register, Department of Justice, Office of Attorney General, 28 CRF Part 36. Copies available from Office on the Americans with Disabilities Act, U.S. Department of Justice, Washington, DC 20530. Phone information: 1-202-514-0301.

13. Bruce Fader, *Industrial Noise Control.* Copyright © 1981 and published by John Wiley & Sons, Inc., New York.

14. Owen J. McAteer, *Electrostatic Discharge Control.* Copyright © 1990 by Owen J. McAteer. Information obtained by McGraw-Hill, Inc., New York. Printed and bound by R. R. Donnelly & Sons Company.

15. Stephen L. Stepkin and Ralph E. Mosely, editors, *Noise Control, A Guide for Workers and Employers.* Copyright © 1984 and published by American Society of Safety Engineers.

16. *Americans with Disabilities Act Accessibility Guidelines Checklist for Buildings and Facilities* (ADAAG Checklist). Prepared by U.S. Architectural & Transportation Barriers Compliance Board (Access Board). Copies available from USATBCB, 1331 F Street NW, suite 1000, Washington, DC 20004. Phone information: 1-800-872-2253.

17. David A. Harris, Alvin E. Palmer, Susan M. Lewis, David L. Munson, Gershon Meckler, and Ralph Gerdes. Second edition edited by David A. Harris, Byron W. Engen, and William E. Fitch. *Planning and Designing the Office Environment*, 2d ed. Copyright © 1991 and published by Van Nostrand Reinhold, New York.

18. Carol Gelderman, *Henry Ford: The Wayward Capitalist*, paperback. Copyright © 1981 by Carol Gelderman, St. Martin's Press, New York.

19. *The Americans with Disabilities Act Checklist for Readily Achievable Barrier Removal.* Copyright © and published by Adaptive Environments Center, Inc. and Barrier Free Environments, Inc. Copies available from Disability and Business Technical Assistance Center. Phone information: 1-800-949-4232.

20. Regional Disability and Business Technical Assistance Center. When you call the phone number shown below, you will be directed to the center nearest to you. Phone information: 1-800-949-4232.

21. AutoCAD LT, computer-aided design program. Available from Autodesk, Inc., 2320 Marinship Way, Sausalito, CA 94965. Phone information: 1-800-228-3601.

Index

Diskette and Book Ordering Information

The price of the diskette is $35.00 and includes handling, postage, and local taxes. The supplier is:

Rocky Mountain Softrak, Inc.
1780 55th Street
Boulder, Colorado 80301
   Telephone number: (800) 999-3475
   Fax number: (303) 444-7179

Please do not call the supplier for the author's address or phone number, since the company will not supply that information. Also, no one there can answer questions about the files on the diskette. The supplier also sells the book *Plant Layout and Flow Improvement*. Call for pricing.

The diskette (copyright by Jay Cedarleaf, 1993) is supplied for MS-DOS or Windows computers and cannot be used on a Macintosh computer. The easiest and least expensive way to order the diskette is by mail, using a copy of the order form and a personal check. Payment can also be made by credit card. The diskette will be shipped with an invoice which you can submit to the petty cash clerk at your company for a refund.

Purchase orders must be on company letterhead and be signed by the buyer. The information requested on the order form must be supplied before your order can be filled. Overseas orders must be paid by cash or credit card in U.S. dollars. When you order a diskette you will receive these items:

1. A labeled, green diskette with the MS-DOS (Windows) files listed in Fig. 1.3, List of Files on Diskette. It can be ordered in either a $5\frac{1}{4}$-in (1.2-MB) or a $3\frac{1}{2}$-in (1.44-MB) format.

2. A plastic pocket that will hold the diskette firmly. It is self-adhesive and can be attached to the inside of the back cover of your book.

3. An invoice for the diskette, but no instructions for its use. The instructions are in the book.

Every effort has been made to keep the price of the diskette reasonable so all readers can increase their productivity by using it. In most cases, the files have been tested by the program manufacturer and found to be satisfactory. If you cannot load the files successfully into your program, return the diskette, postpaid, with an explanation of your problem, to the supplier for a refund. Defective diskettes will be replaced by the supplier at no charge. If you have any comments regarding the book or the diskette files, please write to the author in care of the publisher.

Diskette Order Form

Mail to:

Rocky Mountain Softrak, Inc.
1780 55th Street
Boulder, Colorado 80301
    Fax number: (303) 444-7179

Name _____

Company _____

Address _____

City & State _____

Type of diskette ($35 each):   5¼" (1.2 MB) _____

                                 3½" (1.44 MB) _____

Book: *Plant Layout and Flow Improvement* _____

Check enclosed for _____

Your computer manufacturer _____

Your CAD program _____

Charge to Mastercard or Visa:

Account no. _____

Card expires _____ month _____ year _____

Signature _____